Hydraulic Power Plants: A Textbook for Engineering Students

Authored by

Jafar Mehdi Hassan
Mechanical Engineering Department
University of Technology
Iraq

Salman Hussien Omran
Ministry of Higher Education & Scientific Research / Private Higher Education Directorate
Iraq

Laith Jaafer Habeeb
Mechanical Engineering, University of Technology
Iraq

Alamaslamani Ammar Fadhil Shnawa
Polytechnica University of Bucharest
Iraq

Adrian Ciocănea
Faculty of Power Engineering, Politehnica University of Bucharest
Romania

Hydraulic Power Plants: A Textbook for Engineering Students

Authors: Jafar Mehdi Hassan, Salman Hussien Omran, Laith Jaafer Habeeb,
Alamaslamani Ammar Fadhil Shnawa and Adrian Ciocănea

ISBN (Online): 978-981-14-9412-3

ISBN (Print): 978-981-14-9410-9

ISBN (Paperback): 978-981-14-9411-6

Bentham Science Publishers Pte. Ltd.
80 Robinson Road #02-00
Singapore 068898
Singapore
Email: subscriptions@benthamscience.net

CONTENTS

FOREWORD i
PREFACE iii
CHAPTER 1 THE MOMENTUM EQUATION AND ITS APPLICATION 1
1.1. MOMENTUM AND FLUID DYNAMIC FORCE 1
1.2. APPLICATION OF MOMENTUM EQUATION 4
1.2.1. Force Exerted by the Fluid Jet on a Flat Plate 4
1.2.2. Velocity Diagram (General) 8
1.3. BERNOULLI'S EQUATION FOR RELATIVE MOTION 11
SOLVED PROBLEMS 16

CHAPTER 2 IMPULSE WATER TURBINE 21
2.1. COMPONENTS OF THE PELTON TURBINE 21
2.2. THEORY OF PELTON TURBINE 22
2.2.1. Step by Step Examples 26
2.3. POWER REGULATION MECHANISMS 32
2.4. SUPPLY AND DISCHARGE SYSTEM 33
SOLVED PROBLEMS 37

CHAPTER 3 REACTION TURBINES 49
3.1. TYPE OF REACTION TURBINE 49
3.2. CONSTRUCTION OF REACTION TURBINE 50
3.3. THEORY OF REACTION TURBINE 52
3.4. EFFICIENCY OF REACTION TURBINES 53
3.5. FLOW-RATE THROUGH REACTION TURBINE 54
3.6. VELOCITY TRIANGLE FOR REACTION TURBINE 56
3.7. DRAFT TUBE 57
3.8. NET HEAD 59
3.9. WORKING PROPERTIES OF REACTION TURBINES 63
3.10. POWER REGULATING MECHANISMS 74
3.11. SUPPLY AND DISCHARGE SYSTEMS 79
SOLVED PROBLEMS 87

CHAPTER 4 SIMILARITY LAWS FOR TURBINE SPECIFIC SPEED AND CAVITATIONS 103
4.1. SIMILARITY LAWS 103
4.2. CAVITATION IN TURBINES 107
4.3. TURBINE SELECTION 114
4.4. MARKING TYPES OF TURBINE 115
4.5. HYDRAULIC TURBINES CLASSIFICATION AND SELECTION 118

CHAPTER 5 CENTRIFUGAL AND POSITIVE DISPLACEMENT PUMPS 124
5.1. CENTRIFUGAL PUMPS 124
5.2. CLASSIFICATION AND STRUCTURE OF THE CENTRIFUGAL PUMPS 125
5.3. THEORY OF CENTRIFUGAL PUMPS 131
5.4. HEAD OF THE CENTRIFUGAL PUMPS 134

5.5. FORCE AND POWER OF CENTRIFUGAL PUMPS 136
5.6. EFFICIENCIES OF CENTRIFUGAL PUMPS 136
5.7. FLOW RATE THROUGHOUT THE CENTRIFUGAL PUMP 138
5.8. NET POSITIVE SUCTION HEAD (NPSH) 139
5.8.1. Net Positive Suction Head Available (NPSHa) 139
5.8.2. Net Positive Suction Head Required (NPSHr) 140
5.8.3. Negative Suction Lift 142
5.9. CAVITATION IN PUMP 144
5.10. SIMILARITY LAWS 146
5.11. CHARACTERISTIC CURVES FOR VARIOUS WORKING CONDITIONS 148
5.12. PUMP SELECTION AND PERFORMANCE CHARTS 158
5.13. PUMP AS TURBINE 160
5.13.1. Pump Turbine Classification and Selection 161
SOLVED PROBLEMS 163
5.14. POSITIVE DISPLACEMENT PUMPS 174
5.14.1. Reciprocating Pumps Classification 174
5.14.2. Theory of Reciprocating Pumps 175
5.14.3. Characteristics of Reciprocating Pumps 177
5.14.4. Air Vessel 182
5.15. ROTARY POSITIVE DISPLACEMENT PUMPS 185
5.15.1. Gear Pumps 186
5.15.2. Screw Pumps 190
5.15.3. Vane Pumps 193
5.15.4. Axial Piston Pumps 195
5.15.5. Radial Piston Pumps 198
5.15.6. General Pumping Formulas 200
SOLVED PROBLEMS 203

REFERENCES 211

SUBJECT INDEX 212

FOREWORD

The purpose of hydroelectric power plant is to harness power from water flowing under pressure. As such, it incorporates a number of water driven prime movers known as water turbines.

Hydroelectric power can be developed wherever water continuously flowing under pressure is available. Dams constructed across flowing rivers divert the riverine bounty through the turbine giving rise to such useful power. However, this is not all. Water that is collected in natural or artificial lakes in the high and huge mountains, due to heavy monsoon rain, can be led down to turbines through large pipes known as penstocks.

Basic concepts of hydro power plants and fluid flow are essential in all the engineering disciplines to get better understanding of the course in professional programmers, and obviously its importance as a core subject that needs not to be overemphasized.

The author with his collaborators has been teaching the subject of hydraulic power plants for the past several years, and this monograph is essentially based on the lectures delivered by him. The lecture notes were prepared as; the author comprehend that there was no text, which could provide a coherent readily intelligible account and concise exposition of the subject.

From the experience gained through useful class discussions and feedback, the notes were revised to improve the clarity and necessary explanatory notes where added during each teaching semester. The subject matter has thus been thoroughly tested.

This book is a compilation as no claim is made of its originality. Acknowledgement is due and hereby made to all the authors whose work has been used in the preparation of this text.

Finally, this book emphasizes the need of young engineers acquiring great efficiency in using the tool of study and designing each part of hydraulic power plants such as turbines, pumps, penstocks and other parts in a simple way.

Riyadh S. Al-Turaihi
Mechanical Engineering Department
College of Engineering/Department of Mechanical Engineering
Babylon University
Babil
Iraq

PREFACE

Hydraulic machines-hydraulic turbine, pumps (water and oil types) and reversible hydraulic machines (pump-turbine) are applied in hydroelectric power plants, and water supply system as well as in thermal nuclear and pumped-storage station. In addition, pumps are widely used in the construction of hydraulic structures, such as dams, canals, river and sea ports.

The book describes the construction of hydraulic power plants and treats the theory of the working process for each part, *i.e.* the kinematic and dynamic of the liquid flowing through hydraulic machine and systems, only in the scope necessary for understanding their operation conditions and basic calculation relationships.

The book contains a large number of drawings and charts. It also includes the most important specification and working examples and solved problems, which can be applied in designing and maintenance of hydroelectric power plants, pumping stations and pump installation.

CONSENT FOR PUBLICATION

Not applicable.

CONFLICT OF INTEREST

The author declares no conflict of interest, financial or otherwise.

ACKNOWLEDGEMENTS

Declared none.

Jafar Mehdi Hassan
Mechanical Engineering Department
University of Technology
Iraq

CHAPTER 1

The Momentum Equation and Its Application

Abstract: The major encountered in the hydraulic machine is to find the power developed (or consumed) by (or in) a particular machine. A turbine produces power while pumps, compressors, and fans consume power to run. The power is determined from the dynamic force or forces which are being exerted by the flowing fluid on the boundaries of the flow passage and which are due to the change of momentum. These are determined by applying "Newton's second law of motion".

In this chapter, we present the momentum equation and fluid dynamics forces in a simple way and a step by step manner related to the types of prime movers, turbines, pumps, water wheels ....*etc*. The force and power calculations using the velocity diagram for each type are presented too.

The derivation of Bernoulli's equation for relative motion based on consideration of momentum is very useful to present fluid motion inside a turbine runner or pump impeller of a centrifugal pump, which is shown in this chapter with solved problems.

Keywords: Bernoulli's equation, Momentum equation, Velocity diagram.

1.1. MOMENTUM AND FLUID DYNAMIC FORCE

Network's second law for a system Eq. (1.1) is used as a basis for determining the control volume form of the linear momentum equation. The linear momentum of a system is the product of its mass and velocity. Let **m** be the mass of fluid with velocity **V** as follows:

$$\sum F = \frac{d(mv)}{dt} \quad \textbf{(1.1)}$$

According to that, the general equation across a control volume becomes

$$\frac{\partial}{\partial t}\int_{\text{cv}} \rho v dV + \int_{\text{cs}} \text{v}\rho v dA \quad \textbf{(1.2)}$$

Where cv: control volume

cs: control surface

Jafar Mehdi Hassan, Salman Hussien Omran, Laith Jaafer Habeeb, Alamaslamani Ammar Fadhil Shnawa & Adrian Ciocănea

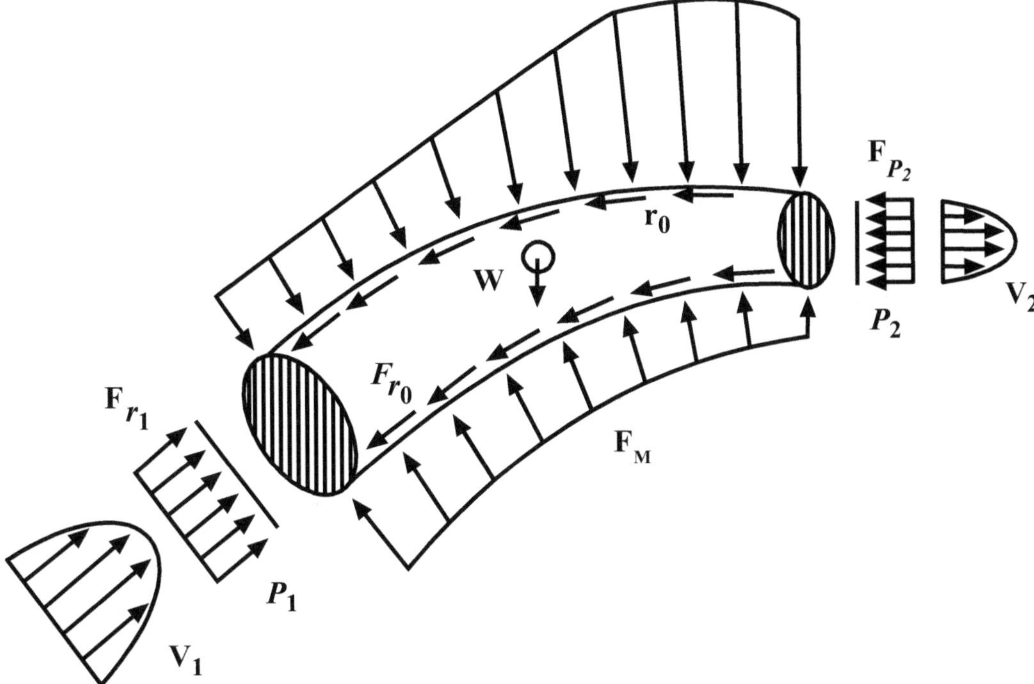

Fig. (1.1). Control volume for flow through a pipe.

The flow is assumed to be steady and the resultant forces and flow parameters are established on the control volume and consist of both resultant forces and equivalent momentum exchanges at the inlet and the outlet (**Fig.1.1**). Therefore, the vector sum of the real applied external force is defined as follows [1]:

$$\sum F = W + F_{p_1} + F_{p_2} + F_{\tau_0} + F_N \tag{1.3}$$

W – Is the weight force. The control volume fluid has a weight acting in the direction of gravity.

$F_{p_1}.F_{p_2}$ – The fluid pressure at inlet and outlet creates a pressure force on each face= PA.

$F_{\tau_o} + F_N$ – The shear stress and the normal stress at the wall or control surface is primarily responsible for maintaining the geometry of the flow field. The stresses are exceedingly difficult to separate therefore, they are lumped together at this point into a **resultant or reaction force** vector **F**, which will act at the center of gravity of the control volume.

The direction and intensity of **F** typically depend on the application. *i.e.*

For Pumps

F- The force exerted by the boundary on the fluid (resultant force) *i.e.* positive.

For Turbine

F- The force exerted by the fluid on the boundary (reaction force (R)) *i.e.* negative

The momentum exchange M_1 and M_2 at the inlet and outlet, respectively, must be analyzed. For steady flow, the right–hand portion of Eq. (1.2) is written as:

$$M_1 + M_2 = -(\rho\overrightarrow{v_1}A_1)\overrightarrow{v_1} + (\rho\overrightarrow{v_2}A_1)\overrightarrow{v_2} \quad \textbf{(1.4)}$$

Or

$$M_1 + M_2 = -\rho Q\overrightarrow{v_1} + \rho Q\overrightarrow{v_2}(A_1)\overrightarrow{v_2} \quad \textbf{(1.5)}$$

When the velocity at the control surface is perpendicular to the area and the velocity is uniform across the respective area.

The minus sign indicates that the momentum is entering the control volume.

The final form at the steady control volume form of "Newton's second law" is

$$W + F_{p_1} + F_{p_2} + F = M_1 + M_2$$

Or

$$\sum F = M_1 + M_2 \quad \textbf{(1.6)}$$

$$\sum F = \dot{m}(\vec{V}_{out} - \vec{V}_{in}) \quad \textbf{(1.7)}$$

This equation is important in the study of turbomachine as it enables to determine the force developed by a fluid machine.

1.2. APPLICATION OF MOMENTUM EQUATION

There are two kinds of application of linear – momentum equation:

a) To determine the force exerted by the flowing fluid on the boundaries of flow passages due to the change of momentum.
b) To determine the flow characteristics when there is some loss of unknown energy in the flow system such as sudden enlargement of a pipe cross-section and hydraulic jump in open channel flow.
In this book we are concerned with the application under (a) which are related to hydraulic machines such as pumps and turbines.

1.2.1. Force Exerted by the Fluid Jet on a Flat Plate

(a) Stationary flat plate (Fig. **1.2**):

A fluid from the nozzle of the jet strikes a flat plate with a velocity V. The plate is stationary and perpendicular to the centerline of the jet. Then

$$F = \dot{m}(\vec{V}_{out} - \vec{V}_{in}) \quad \textbf{(1.8)}$$

$$\vec{V}_{out} = V_2 = 0 \quad Stationary\ plate$$

$$\vec{V}_{in} = V_1$$

$$\therefore F = \dot{m}(0 - \vec{V}_{in})$$

$$F = -\dot{m}\vec{V}_{in}$$

$$since\ \dot{m} = \rho Q$$

And $Q = AV$

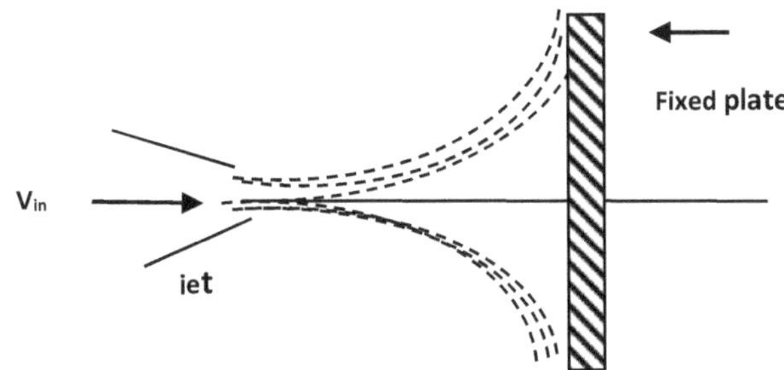

Fig. (1.2). Flow direction on the vertical flat plate.

Then

$$F = -\rho Q\vec{V}_{in} \quad \textbf{(1.9)}$$

The force exerted by the fluid on the plate is reaction force. According to ' Newton's Law of action and reaction, this will be equal and opposite *i.e*:

$$\therefore R = \rho AV^2 \quad \textbf{(1.10)}$$

$A = cross - sectional\ area\ of\ the\ jet.$

(b) Inclined Stationary flat plate (Fig. **1.3**):

Under the similar conditions, the dynamic force acting on the plate is given by:

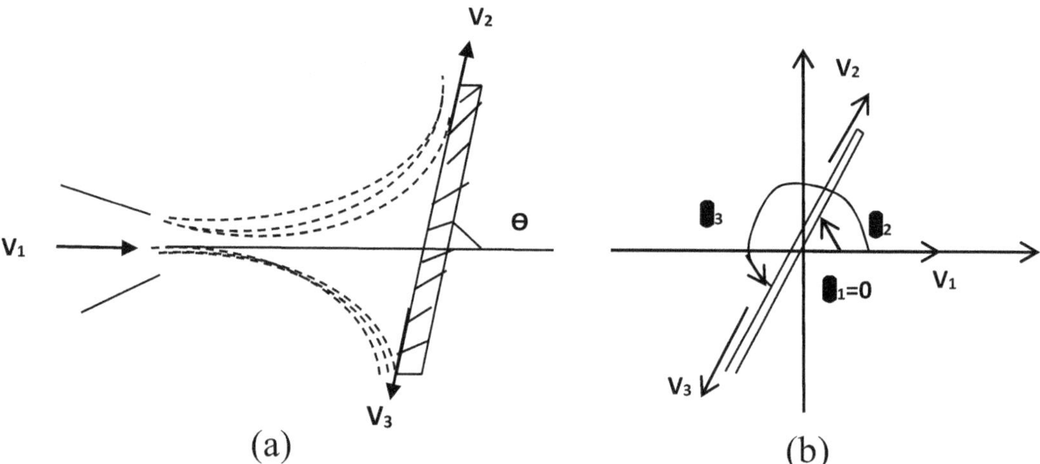

Fig. (1.3). (**a**) Flow direction on an inclined flat plate (**b**) The angles of flow direction.

$$F = \dot{m}(\vec{V}_{out} - \vec{V}_{in})$$

$$\sum F = \sum \dot{m}\vec{V}_{out} - \sum \dot{m}\vec{V}_{in} \quad \textbf{(1.11)}$$

$$\sum F_x = \rho Q_2 V_2 \cos\theta_2 + \rho Q_3 V_3 \cos\theta_3 - \rho Q_1 V_1 \cos\theta_1 \quad \textbf{(1.12)}$$

$$\sum F_y = \rho Q_2 V_2 \sin\theta_2 + \rho Q_3 V_3 \sin\theta_3 - \rho Q_1 V_1 \sin\theta_1 \quad \textbf{(1.12a)}$$

Where Q_2 and Q_3 are the flow rates moving upward and downward along the plate respectively.

And $Q = AV_r$

The direction of the reaction vectors F_x and F_y is not known, therefore directions for these will be assumed positive, and the angle $\theta_1.\theta_2$ and θ_3 will be used to describe the flow direction (Fig. **1.3b**). If the assumption is incorrect the values calculated for F_x and F_y will be negative suggesting that the original direction needs to be reversed. The magnitudes of F_x and F_y remain the same in either case.

(c) Moving plate (Fig. **1.4**):

Let the plate (Fig. **1.2**) move with a velocity U in the same direction of the jet, after the jet with velocity V has struck the plate, then the change in velocity is $(V - U)$. Also, the quantity of water striking the plate per second is given by cross-sectional area that is multiplied by the velocity of the jet relative to the plate which is V_r then

For moving plate $V \rightarrow V_r$

V_r = Relative velocity = $V - U$ and $Q = AV_r$

U = Velocity of the plate.

$in\ x - direction$

$\vec{V}_{out} = 0$, $\theta_2 = 90°$ (normal to the paper $V \cos\theta$) and $V_{in} = V_{r1}$ $\theta_1 = 0$

$\therefore\ F = -\dot{m}\vec{V}_{r1}$

And the reaction force becomes

$R = \dot{m}\vec{V}_{r1}$ $\dot{m} = rA(V - U)$ For single moving plate

Or $\therefore R = rA(V - U)^2$ **(1.13)**

For series moving plate the distance between plate and nozzle is constantly increasing by $U\ m/s$. A single moving plate is, therefore, not a practical case. If, however a series of plates (Fig. **1.4**) so arranged that each plate appeared successively before the jet in the same position and always moving with a velocity

U in the direction of the jet, then the whole flow from the nozzle is utilized by the plate. Thus, the weight of water striking the plate would be

$\dot{m} = rAV \quad and$

$\vec{V}_{in} = V_{r1} = V - U \quad \theta = 0°$

$\vec{V}_{out} = 0 \qquad \theta_2 = 90°$

$$\therefore F = -r\,AV(V - U) \tag{1.14a}$$

Or the reaction force becomes

$$R = r\,AV(V - U) \tag{1.14b}$$

then

$Power = F.U$ Series of moving plates (vanes)

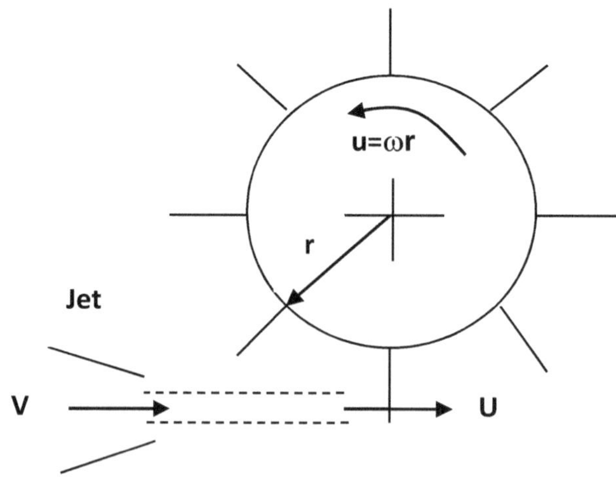

Fig. (1.4). Flow direction on a series of moving plates.

1.2.2. Velocity Diagram (General)

The jet of water impinges on a moving curved plate with a velocity V_1 at the inlet with an angle α_1 (Fig. **1.5a**) with respect to x-direction. The curvature of the plate at the point where the jet strikes may or may not make the same angle α_1 with x-direction. Let the angle of curvature of the plate at the inlet with the reversed direction of motion of the plate be β_1. As soon as the jet falls over the plates, the velocity of the jet becomes relative to the motion of the plate. This velocity is denoted by w_1. Its direction will be tangential to the point of the inlet.

Let the velocity of the jet relative to the plate motion at the outlet be denoted by w_2 and inclined at the outlet by an angle β_2.

Now, the absolute velocity of water at inlet and outlet V_1 and V_2 will be the vector sum of the two velocities:-

$$\vec{V} = \vec{w} + \vec{U}$$

The magnitude and direction of absolute velocities is determined by applying the Law of Parallelogram of Force (Fig. **1.5a**).

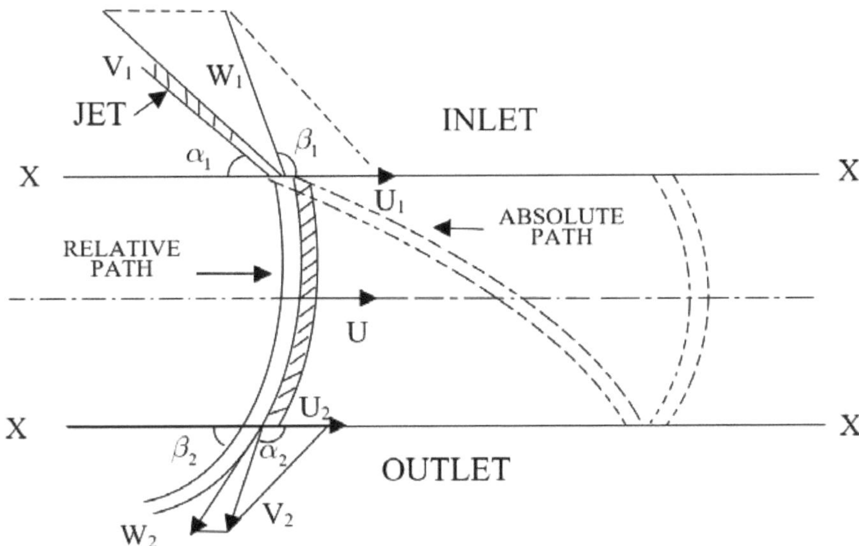

Fig. (1.5a). Jet Falling on Moving curved Plate with Obtuse Discharge Angle α_2.

The direction of the absolute velocity of water at any point will be tangential to the absolute path of water. Similarly, the direction of relative velocity of the water at any point will be tangent to the relative path of water. The direction of the peripheral velocity of the plate is always horizontal. With the direction of all the three velocities U, V and w being known, the velocity triangle can be drawn at any point of the path. The velocity triangles have been shown 1, 4 and 6 in Fig. (**1.5b**).

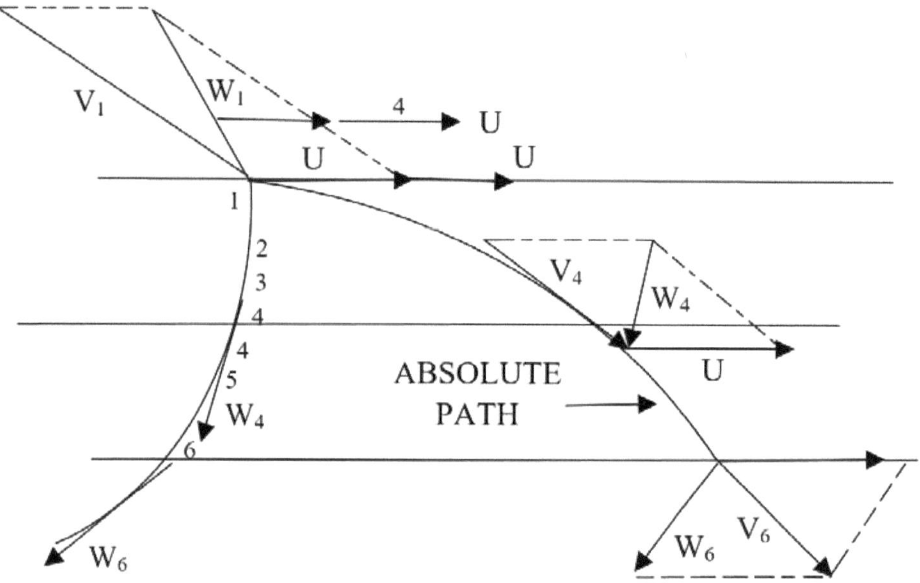

Fig. (1.5b). Determination of Absolute Path of water Particle from its Relative Path.

Fig. (**1.5c**) shows the Typical Velocity Triangles or diagrams and Pumps Blades. They have been taken and are drawn similar to Fig. (**1.5b**).

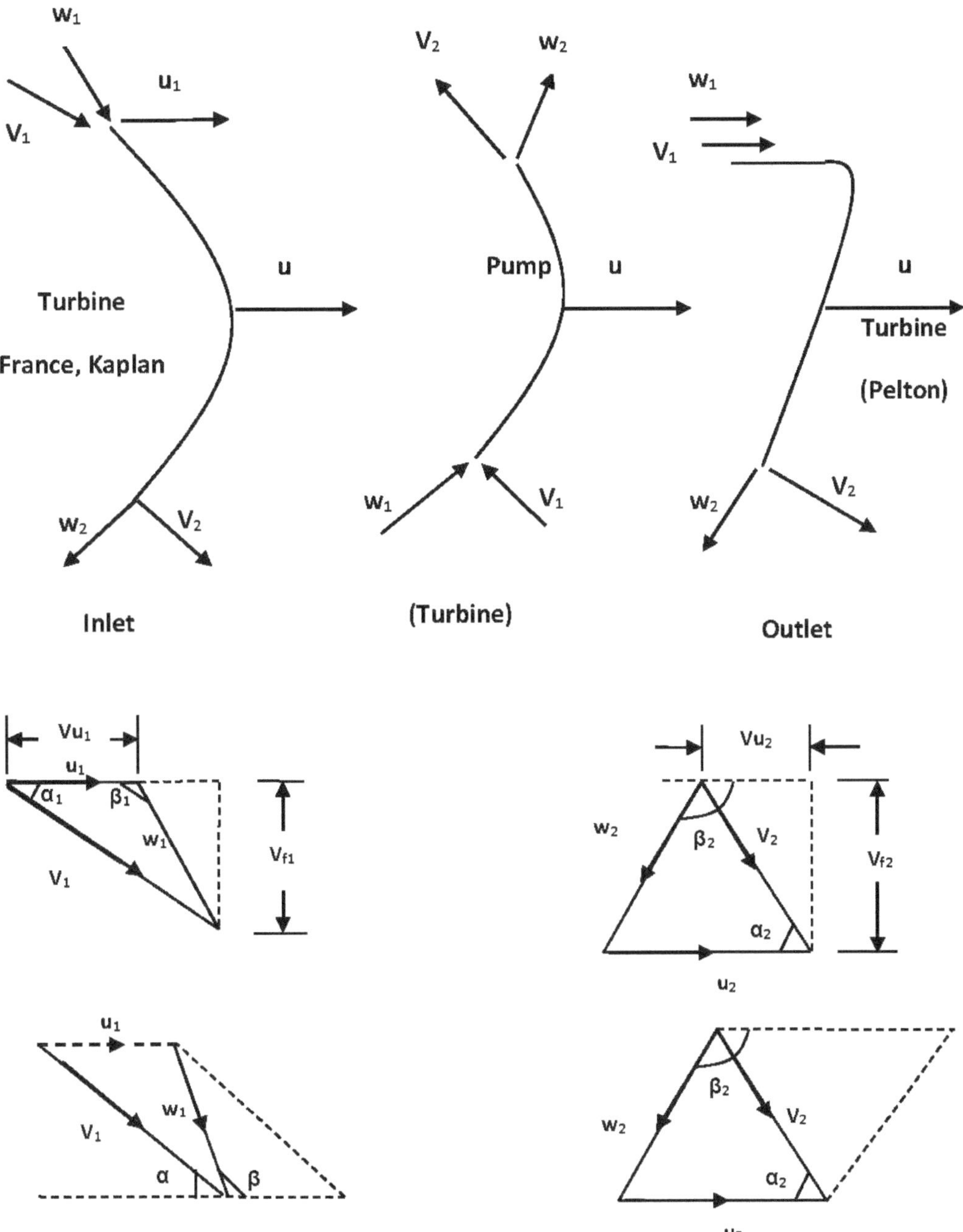

Fig. (1.5c). Typical Velocity triangle for Flow Over Turbine and Pump Blades.

V_1, V_2 : The inlet and outlet velocity (absolute).

$\alpha_1, \propto_2$: The angles of the inlet and outlet velocities.

U_1, U_2 : The velocities of the (Vane, blade, tangential, peripheral) wheel.

$$U_1 = U_2 = U \qquad \text{For impulse turbine.}$$

W_1, W_2 : relative velocities at inlet and outlet. It is tangential to the blade at the inlet and outlet.

β_1, β_2 : The (blade, vane, wheel, relative velocities) angles at inlet and outlet.

$V_u = V \cos \alpha$(Whirl velocity) at inlet and outlet.

$V_f = V \sin \alpha$(Flow velocity) at inlet and outlet.

1.3. BERNOULLI'S EQUATION FOR RELATIVE MOTION

Consider the motion of fluid inside a turbine runner or an impeller of centrifugal pump (Fig. **1.6**). Let its angular velocity ω be constant and the relative velocity be W. Consider a small element of fluid of length ds breadth b and thickness l and l+dl at two ends which is a part of the stream tube.

Forces on the element are fluid pressure, gravitational, and centrifugal force.

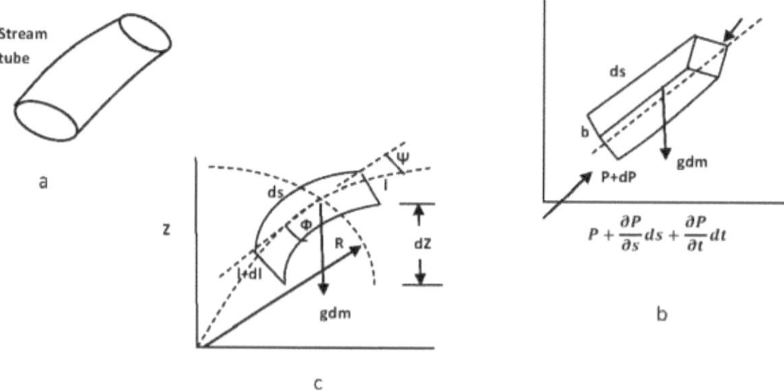

Fig. (1.6). Absolute Motion of Fluid Particles (**a**) Stream tube (**b**) Pressure variation (**c**) Relative Motion of Fluid Particles.

Φ: the angle between the tangent

and the gravity force

Ψ: the angle between the

tangent and the radians of curvature.

For steady stale $\frac{\partial}{\partial t} = 0$

Acceleration in the direction of relative velocity W

$$\frac{dw}{dt} = \frac{\partial w}{\partial t} + w\frac{\partial w}{\partial s} = w\frac{\partial w}{\partial s} \qquad \text{for steady flow.} \tag{1.15}$$

Mass of fluid acceleration $= \frac{\gamma}{g}(blds)$

Forces acting on the element are:

a) Weight, (gravitational force) $= \gamma blds$

b) Centrifugal force $= \frac{\gamma}{g} blds.Rw^2$

c) Pressure difference force $= Pdl - (P + dP)b(l + dl)$

$$= -bldP \qquad (neglecting\ \frac{dl}{l}\ terms)$$

Now resultant force in the direction of streamline =mass ×acceleration

$$\therefore \gamma blds\cos\emptyset - bldP - \frac{\gamma}{g} blds.Rw^2 = \frac{\gamma}{g} bldsw\frac{dw}{ds}$$

$$w\frac{dw}{ds} = ds\cos\emptyset - \frac{dP}{\gamma} - \frac{Rw^2}{g} ds\cos\emptyset$$

Substituting $ds\cos\emptyset = -dZ\ and\ ds\cos\varphi = -dR$

$$\therefore \frac{wdw}{g} = -dZ - \frac{dP}{\gamma} + \frac{R\rho^2}{g}dR \quad \textbf{(1.16)}$$

$$or\ w.\frac{dw}{g} + dZ + \frac{dP}{\gamma} - \frac{R^2\rho^2}{2g}dR = 0$$

$$\frac{w^2}{2g} + Z + \frac{P}{\gamma} - \frac{R^2\rho^2}{2g} = C \qquad since\ U = R\rho$$

$$\therefore \frac{w^2}{2g} - \frac{U^2}{2g} + \frac{P}{\gamma} + Z = C \qquad \text{B.E for relative motion} \quad \textbf{(1.17)}$$

$$\therefore \frac{{w_1}^2}{2g} - \frac{{U_1}^2}{2g} + \frac{P_1}{\gamma} + Z_1 = \frac{{w_2}^2}{2g} - \frac{{U_2}^2}{2g} + \frac{P_2}{\gamma} + Z_2 + losses_{1-2} \quad \textbf{(1.18)}$$

Ex.1: A jet of water, moving at 60 m/s is deflected by a vane moving at 25 m/s in a direction at 30° to the direction of the jet. The water leaves the blades normally to the motion of the vanes.

Draw inlet and outlet velocity triangle, and find the vane angles at the entry and exit. Take relative velocity at the outlet to be 0.85 of the relative velocity at the inlet

Solution:-

$V_1 = 60\frac{m}{s}$; $U_1 = 25\frac{m}{s}$; $\alpha_1 = 30°$

$w_2 = 0.85w_1$; $\alpha_2 = 90°$

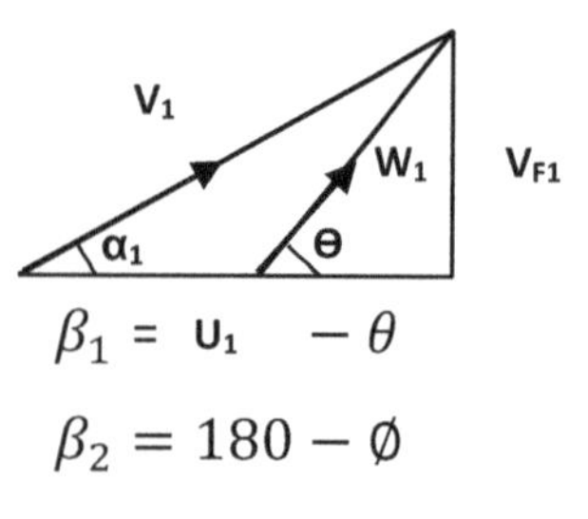

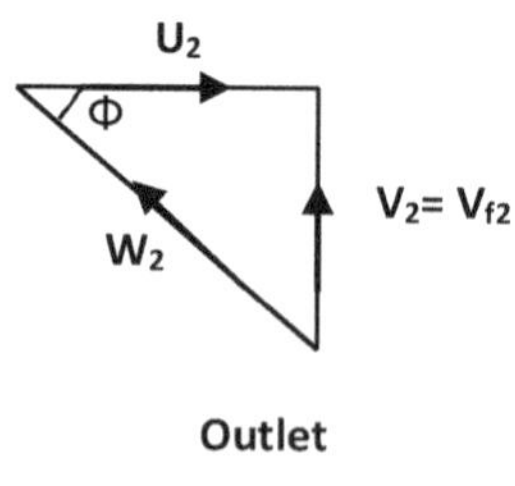

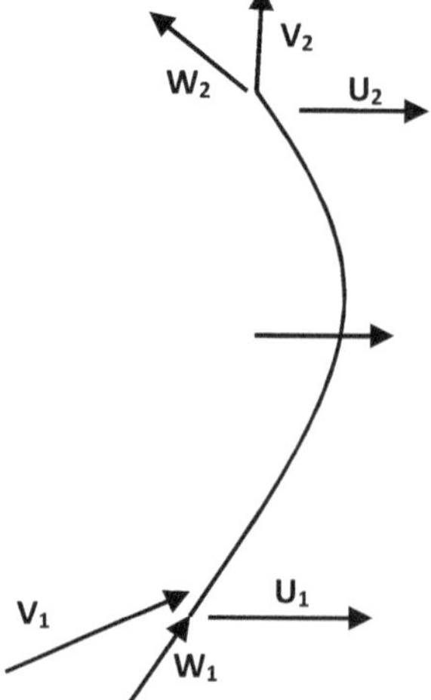

$\beta_2 = 180 - \emptyset$

From inlet velocity triangle

$$V_{u1} = V_1 \cos\alpha_1 = 60 \times 0.86 = 51.96\, m/s$$

$$V_{f1} = V_1 \sin\alpha_1 = 60 \times 0.5 = 30\, m/s$$

$$\tan\theta = \frac{V_{f1}}{V_{u1} - U} = \frac{30}{51.96 - 25}$$

$$\theta = 48.07° \quad \beta_1 = 180 - 48.07 = 132°$$

From inlet velocity triangle

$$w_1 = \frac{V_f}{\sin 48.07} \qquad \therefore\ w_1 = 40.34\, m/s$$

$$\therefore\ w_2 = 0.85 w_1 = 34.29\, m/s$$

$$and\ U_1 = U_2$$

From outlet velocity triangle

$$\cos\emptyset = \frac{U_2}{w_2} = \frac{25}{34.29}$$

$$\emptyset = 43.2 \qquad \beta_2 = 136.8°$$

Ex.2: A jet of water, having a velocity of 30 m/s impinges on a series of vanes with a velocity of 15 m/s. take jet makes an angle of 30° to the direction of motion of vanes when entering and leaves at an angle of 120°. Sketch velocity triangle at inlet and outlet and determine.

a) Angle of the vane at inlet and leaves without stock.

b) Work done per kg of water entering the vanes.

c) Efficiency.

Solution:-

$$V_1 = 30\frac{m}{s}\ ;\ U_1 = U_2 = 15\frac{m}{s}\ ; \alpha_1 = 30°\ ; \alpha_2 = 120°$$

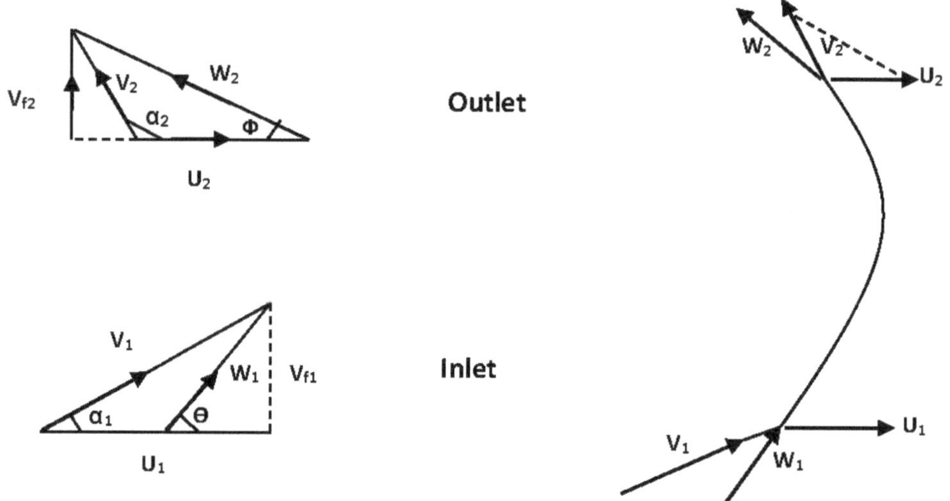

From inlet velocity triangle

$$V_{u1} = V_1 \cos\alpha_1 = 30 \times \cos 30 = 25.98\ m/s$$

$$V_f = V_1 \sin\alpha_1 = 30 \times \sin 30 = 15\ m/s$$

$$\therefore \tan\theta = \frac{V_{f1}}{V_{u1} - U_1} = \frac{15}{25.98 - 15}$$

$$\therefore \theta = 53.8°$$

$$\therefore \beta_1 = 126.2°$$

$$also \quad W_1 = \frac{V_f}{\sin 53.8°} = \frac{15}{\sin 53.8°} = 18.59\ m/s$$

From outlet velocity triangle

$$\frac{U_2}{\sin(60-\emptyset)}=\frac{W_2}{\sin 120^\circ}$$

$$\frac{15}{\sin(60-\emptyset)}=\frac{18.59}{\sin 120^\circ}$$

$$(60-\emptyset)=44.3^\circ \quad \emptyset=15.67^\circ$$

$$\therefore \beta_2=164.3^\circ$$

$$V_{u2}=W_2\cos 15.67-U_2$$

$$=18.59\times 0.963-15=2.9\,m/s$$

$$\therefore \frac{WD}{s} per\ kN\ of\ water=\frac{1}{g}(V_{u1}U_1-V_{u2}U_2)$$

$$=\frac{U}{g}(V_{u1}-V_{u2})=\frac{15(25.98-2.9)}{9.81}$$

$$=35.29\,N.m$$

$$Efficiency=\frac{Work\ done\ per\ kg\ of\ water}{Energy\ of jet\ per\ kg\ of\ water}$$

$$=\frac{35.29}{\frac{V^2}{2g}}$$

$$Efficiency=\frac{35.29}{45.87}=77\%$$

SOLVED PROBLEMS

Q.1. A circular jet delivers water at the rate of 60 l/s with a velocity of 24 m/s. The jet impinges tangentially on a vane moving in the direction of the jet, with a velocity of 12 m/s. The vane angle at outlet 45°, through what angle will if deflect the jet.

Solution:-

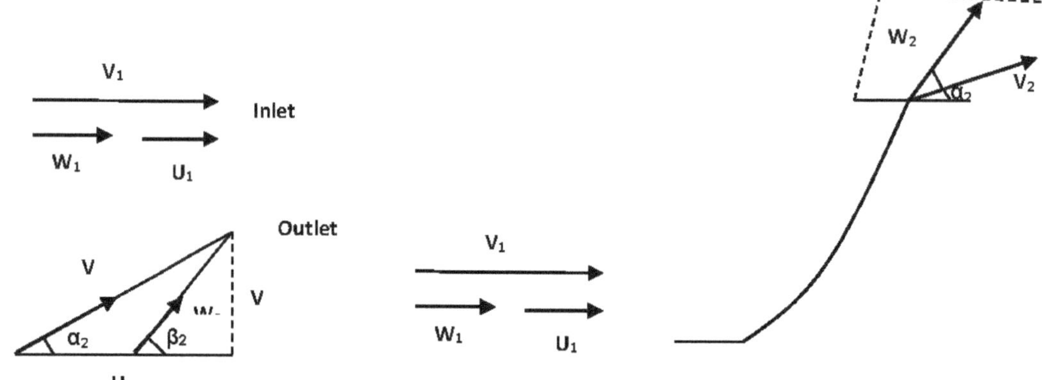

$$Q = 60\frac{l}{s}\ ;\ U_1 = U_2 = 12\frac{m}{s}\ ;\alpha_1 = 0\ ;\beta_2 = 45^\circ$$

$$\alpha_2 = ?\ \ ;\ w_1 = w_2$$

$$V_{u1} = V_1 \cos\alpha_1 = V_1 = 24\ m/s$$

$$w_1 = V_{u1} - U_1 = 12\ m/s$$

From velocity triangle, since $U_2 = w_2 = 12\ m/s$

$$\therefore \alpha_2 = \frac{45}{2} = 22.5^\circ$$

$$Work = \dot{m}(V_{u1}U_1 - V_{u2}U_2) \qquad U_1 = U_2$$

$$= r\ Q(V_{u1} - V_{u2})U$$

$$V_{u2} = U_2 + W_2 \cos 45^\circ = 12 + 12\cos 45^\circ = 20.5\ m/s$$

$$\therefore Work/s = 1000 \times 0.06(24 - 20.5) \times 15$$

$$= 3150\ W$$

Q.2. A jet of water 100 mm in diameter, moving with a velocity of 25 m/s in the direction of the vanes, enters the vane moving with a velocity of 12.5 m/s. If the jet leaves the vane at an angle of 60° with the direction of motion of the vanes, find

a) The force on the vanes in the direction of motion

b) The work done per second (Power)

Solution:-

Diameter of jet $= 100\, mm = 0.1m \qquad a = \frac{\pi}{4}(0.1)^2 = 0.00785\, m^2$

$$V_1 = 25\, m/s\,;\, U_1 = U_2 = 12.5\, m/s$$

$$\alpha_1 = 0\,;\, \alpha_2 = 60°$$

$$F = r\, Q(V_{u1} - V_{u2})$$

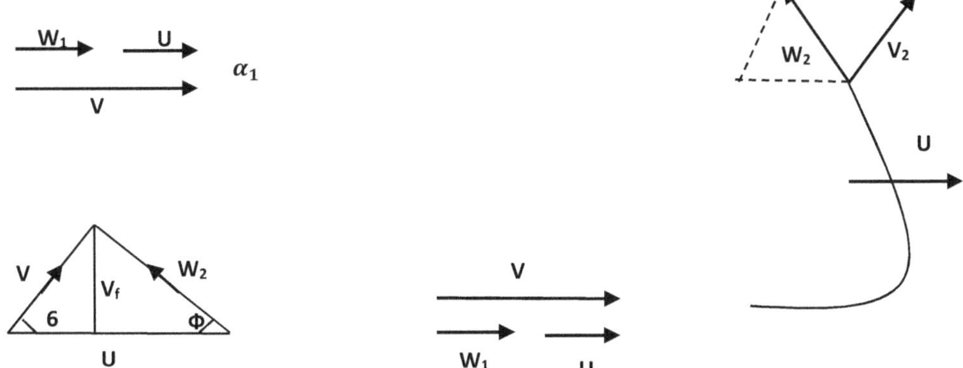

$$V_{u1} = V_1 \cos\alpha_1 = 25\, m/s$$

$$w_1 = V_{u1} - U_1 = 25 - 12.5 = 12.5\, m/s$$

From outlet triangle $w_2 = U_2 \qquad \therefore \emptyset = 60°$

$$\therefore \frac{V_2}{\sin 60°} = \frac{w_2}{\sin 60°}$$

$$\therefore V_2 = w_2 = 12.5$$

$$Then \;\; V_{u2} = V_2 \cos 60° = 6.25 \; m/s$$

$$F = \dot{m}(V_{u1} - V_{u2})U = 1000 \times 0.00785 \times 25(25 - 6.25)$$

$$= 3.68 \; kN$$

$$Power = F.U = 3.68 \; \times 12.5 = 46 \; kW$$

Q.3. A jet of water 50 mm in diameter impinges on a curved vane and is deflected through an angle of 135°. The vane moves in the same direction as that of the jet with a velocity of 5 m/s. if the rate of flow of the water is 30 l/s. Determine,

a) component of force on the vane in the direction of motion.

b) Power developed by the vane.

c) Efficiency.

Solution:-

$$d = 50 \; mm \;\; \therefore a = 0.00196 \; m^2 \; ; \beta_2 = 135°$$

$$\emptyset = 45° \; ; \; U_1 = U_2 = 5 \; m/s \; ; Q = 30 \; l/s$$

$$w_1 = w_2$$

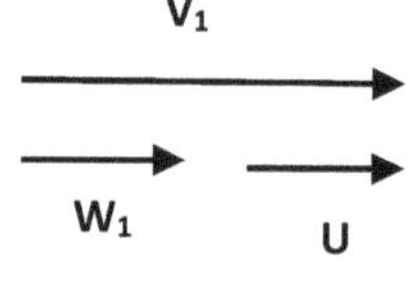

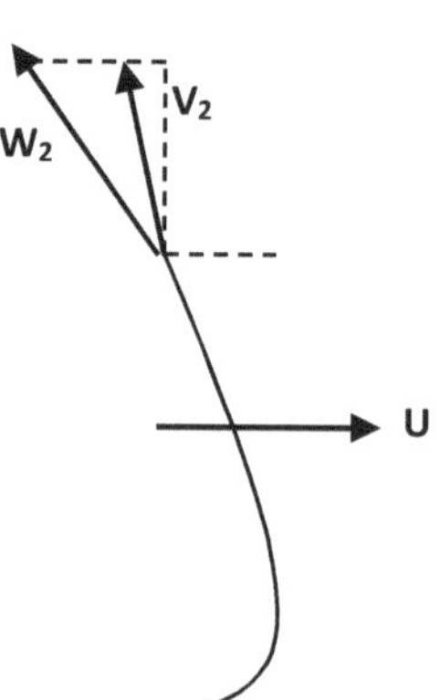

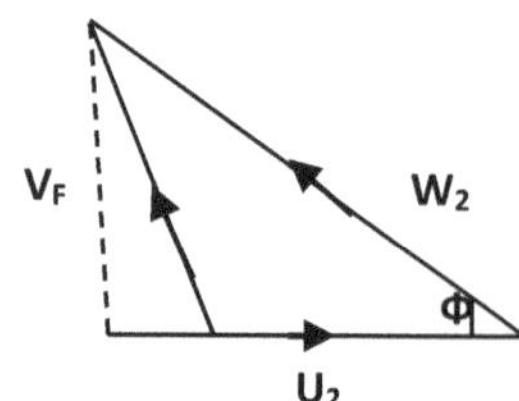

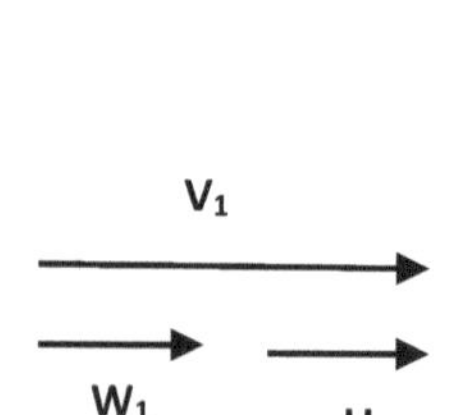

$$V = \frac{Q}{a} = \frac{0.03}{0.00196} = 15.31\ m/s$$

$$V_{u1} = V_1 \cos \alpha_1 = 15.31\ m/s \qquad ;\ \alpha_1 = 0$$

$$w_1 = V_1 - U = 15.31 - 5 = 10.31\ m/s$$

$$V_{u2} = w_2 \cos 45° - U = (10.31 \times 0.707) - 5$$

$$= 2.29\ m/s$$

$$\therefore F = r\ Q(V_{u2} - V_{u1}) = 1000 \times 0.03(15.31 - 2.29)$$

$$= 390.6\ N$$

$$Power = F.U = 390.6 \ \times 5 = 1953\ watt$$

$$also\ \ K.E/s = \frac{1}{2}\dot{m}V^2 = \frac{1}{2}r\ QV^2$$

$$= \frac{1}{2} \times 1000 \times 0.03 \times (15.31)^2 = 3516\ watt$$

$$\eta = \frac{1953}{3516} = 55.5\%$$

CHAPTER 2

Impulse Water Turbine

Abstract: Impulse water turbine (Pelton turbine or Pelton wheel) is the standard type of turbines that are widely used nowadays. It is also called a free jet turbine. Pelton turbine operates under the high head of water and therefore, requires a comparatively less quantity of water.

In this chapter, the components of hydro-electric power plants using this type of turbine are presented with neat sketches. Theory of power, all mathematical calculations using velocity diagrams, solutions for single or multi-sets, power regulation components, supply, and discharge systems are also presented. Finally, solved problems are illustrated in detail at the end of this chapter.

Keywords: Components of impulse turbine, Pelton turbine, Power regulation mechanisms, Supply and discharge system.

2.1. COMPONENTS OF THE PELTON TURBINE

The impulse turbine runs by the impulsive water force (Pelton wheel). It is typically used for high-water head applications within a range of 300 to 1500 m, 0.5 to 20 m^3/s, and 200 MW for water head, flow rate, and net output power, respectively. The main components of the impulse turbine are shown in Fig. (**2.1**) [2].

1- Guide Mechanism: the primary function of the guide mechanism is to control the quantity of water that is passing through the nozzle and striking the buckets. It consists of the nozzle and the governor.

2- Buckets and Runner: each bucket is divided vertically into two parts by a splitter, which is a sharp edge in the center, giving the shape of a double hemispherical cap.

3- Casting: the casting of the Pelton wheel has no hydraulic function to perform. It is necessary only to prevent water from splashing, guide the water to the tailrace, and also works as a safeguard against accidents.

Jafar Mehdi Hassan, Salman Hussien Omran, Laith Jaafer Habeeb, Alamaslamani Ammar Fadhil Shnawa & Adrian Ciocănea

4- Hydraulic Brake: this part consists of a small nozzle fitted in such a way that on being opened, it directs a jet on the back of the buckets to bring the revolving runner quickly to rest.

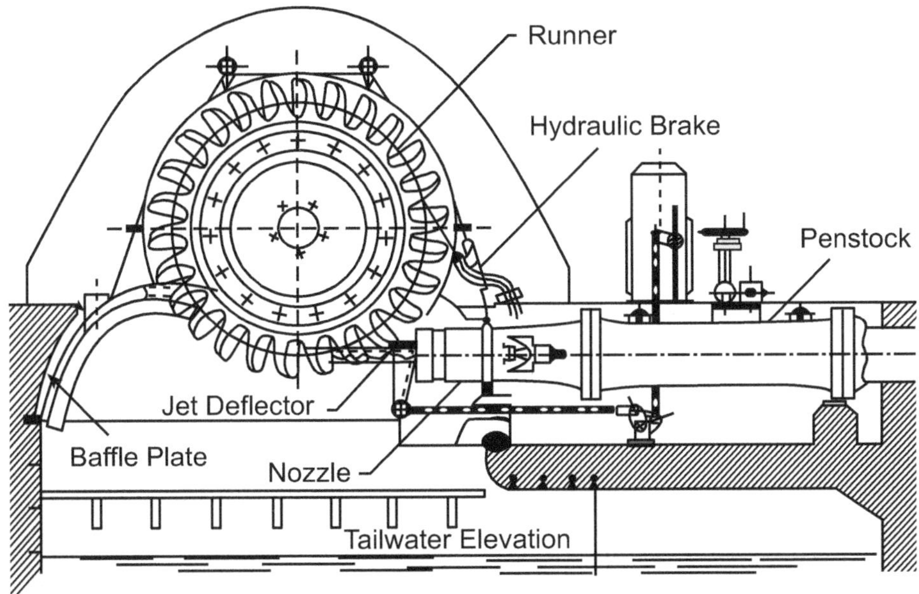

Fig. (2.1). Schematic diagram of Pelton Turbine contraction.

2.2. THEORY OF PELTON TURBINE

1. Turbine Power

Power input can be given by,

$$P_a = \gamma Q H_T \quad kW \tag{2.1}$$

where P_a is the available power measured in kW, H_T is the net head acting at the turbine inlet in m, and Q is the flow rate in m^3/s.

The supplied power by the turbine (kW) can be expressed as,

$$P_t = P_a \eta_t \tag{2.2}$$

where, P_t and η_t are the supplied power and turbine efficiency, respectively.

2. Discharge of the Nozzle and Jet Diameter

Fig. (**2.2**) shows a schematic diagram of the spear, nozzle, and jet components. The discharge depends on the least diameter of the jet that is precisely measured at the vena contraction. The discharge can then be expressed in terms of the diameter using the following relation,

$$Q = \frac{\pi}{4} {d_i}^2 V_1 \quad \textbf{(2.3)}$$

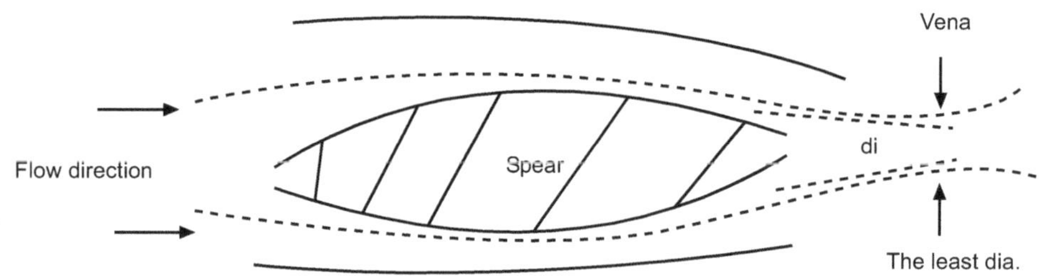

Fig. (2.2). Spear, nozzle, and jet.

d_i = The least diameter of the jet (m)

V_i = Velocity of the jet (m/s)

Q = Discharge (m^3/s)

$$V_i = C_V \sqrt{2gH} \quad \textbf{(2.4)}$$

C_V=Velocity coefficient

The least diameter can be calculated by solving Eqs. (2.3) and (2.4) together so that,

$$\therefore\ d_i = \sqrt{\frac{4Q}{\pi C_V \sqrt{2gH}}} \quad \textbf{(2.5)}$$

3. Multi Jets

The specific speed of Pelton turbine for a single jet and one runner (wheel) can be calculated by,

$$N_s = \frac{N\sqrt{P_t}}{H^{\frac{5}{4}}} \quad (2.6)$$

N_s = Specific speed $N^{\frac{1}{2}}/m^{\frac{3}{4}} \cdot s^{\frac{3}{2}}$

N = speed of the turbine (rpm)

The total power of the turbine depends on the number of the attached nozzles to the turbine. For example, for a single wheel with two attached nozzles (two jets are provided), the total power can be given in terms of nozzles number as,

$$Power = 2P_t \quad (2.7)$$

It is worthy to mention that the maximum number of nozzles that are so far used in a turbine is 6 in the vertical installation and 2 in the horizontal installation.

4. Mean Diameter of Pelton Wheel

Fig. (**2.3**) shows the cross-section of the Pelton wheel. The mean diameter is expressed by the symbol D. For design purposes, the mean diameter can be calculated using the mathematical manipulations described below. The wheel velocity is a function of water head (H) and the coefficient of speed ratio (∅), (*i.e.*, the ratio of tangential velocity U to jet velocity ($V = \emptyset\sqrt{2gH}$), which is typically chosen between 0.44 to 0.46 for good design and, therefore, it can be described by:

$$U = \emptyset\sqrt{2gH} \quad (2.8)$$

Besides, the term U is defined by the wheel revolutions such that $U = \pi DN/60$. Using the definitions above, the least diameter can be obtained by the expression,

$\frac{\pi DN}{60} = \emptyset\sqrt{2gH}$ and for more convenience, the diameter D can be easily expressed by,

$$D = \frac{60\emptyset\sqrt{2g}}{\pi N} \quad (2.9)$$

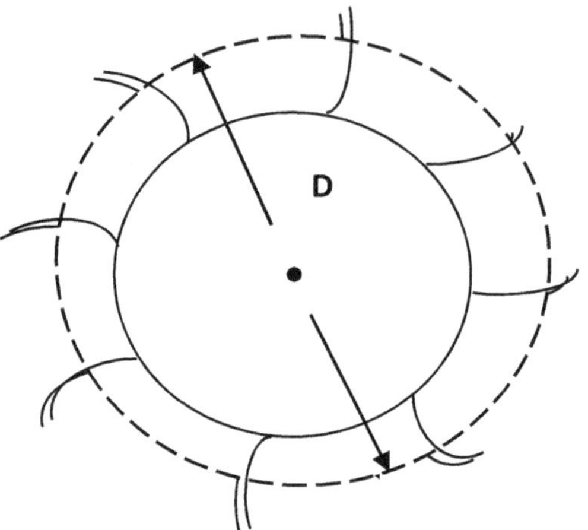

Fig. (2.3). Mean diameter of the Pelton turbine.

5- Jet Ratio

It is an essential feature of the Pelton turbine and it influences its characteristics. It is generally used instead of the specific speed in the selection and design of Pelton Turbines.

$$Jet\ ratio = \frac{mean\ dia.of runner}{least\ dia.of\ the\ jet} \quad \textbf{(2.10)}$$

$$i.e. \quad m = \frac{D}{d_i} in\ general\ (11 \rightarrow 18) \quad \textbf{(2.11)}$$

The jet ratio (m) is another crucial design parameter that should be taken in consideration throughout the turbine design process. Jet ratio can be defined as the ratio between the mean diameter of the runner to the least diameter of the jet and can be given by $m = \frac{D}{d_i}$. The typical values of the jet ratio can be chosen in the range of 11 to 18.

To find the maximum efficiency (η_{max}), the principle of maxima can be easily applied so that the maxima condition must be satisfied as $\frac{d\eta_h}{du} = 0$.

6- Velocity Diagram

Since the rotational speed of the wheel is w, and the tangential speed at the inlet and outlet of the wheel are:

Fig. (**2.4**) shows the ideal velocity diagram. For an ideal Pelton wheel $r_1 = r_2$, therefore, $U_1 = U_2 = U$.

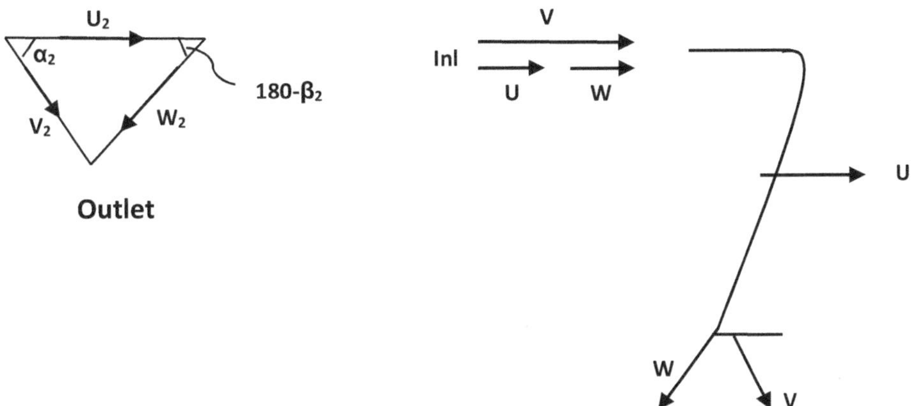

Fig. (2.4). Ideal velocity diagram.

In addition, the relative velocity, when it is opened to the atmosphere, is described by $w_1 = w_2 = V_1 - U$.

2.2.1. Step by Step Examples

Example 1: The mean bucket speed of a Pelton turbine is 14 m/s. The supplied water flow rate by the water jet at the head of 45 m is 800 lit/s. If the jet is deflected by the bucket at an angle of 165°. Find the power and efficiency of the turbine.

Assume $C_V = 0.985$

Solution:-

$$\alpha_1 = 0 \ ; \ U_1 = U_2 = 14\frac{m}{s} \ ; Q = 800\frac{l}{s}$$

$H = 45\,m\ ;\ \beta_2 = 165°\ ; C_V = 0.985$

$V_1 = C_V\sqrt{2gH} = 0.985\sqrt{2 \times 9.81 \times 45}$

$= 29.2\ m/s$

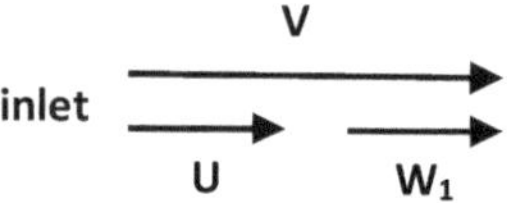

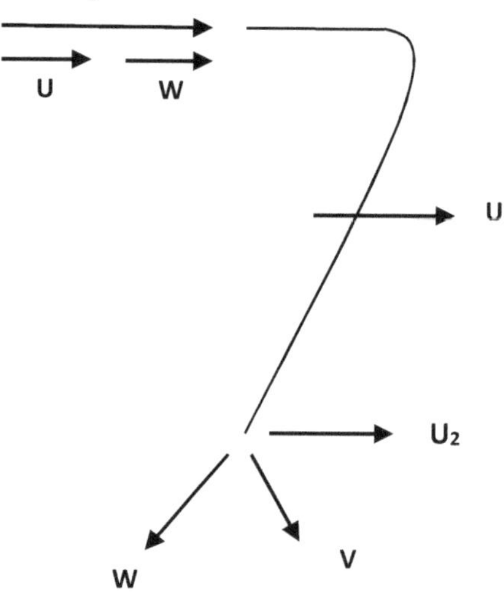

$Pelton\ wheel \begin{cases} U_1 = U_2 = U \\ w_1 = w_2 \end{cases}$

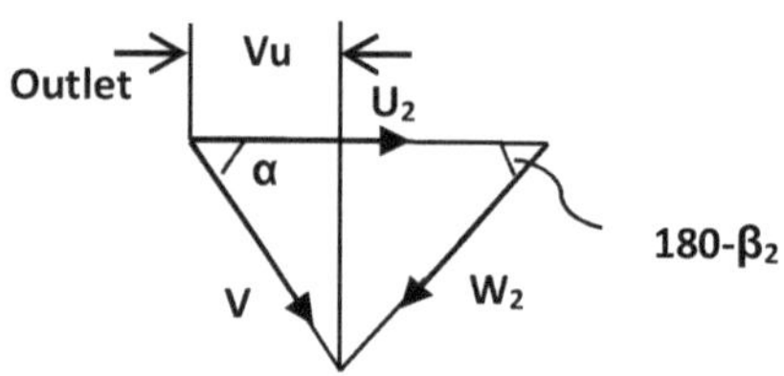

From the inlet velocity triangle,

$V_{u1} = V_1 \cos\alpha_1 = U_1 + w_1 \cos\beta_1$

$\therefore V_1 = U_1 + w_1 \qquad \alpha_1 = 0 \qquad \beta_1 = 0$

$\therefore w_1 = V_1 - U_1 = 29.4 - 14 = 15.2\ m/s = w_2$

From the outlet velocity triangle

$w_2 = 15.2\ m/s \quad U = 14\ m/s$

$\therefore V_{u2} = V_2 \cos\alpha_2 = U - w_2 \cos(180 - \beta_2)$

$= 14 - 15.2 \cos 15°$

$= -0.7\ m/s$

$\therefore Power\ (P_t) = F.U = \rho Q(V_{u1} - V_{u2})U$

$= 1000 \times 0.8 \times (V_1 \times \cos \alpha_1 - V_2 \times \cos \alpha_2) \times U$

$= 1000 \times 0.8 \times (29.2 + 0.7) \times 14$

$= 334.88\ kW$

$$\eta_t = \frac{P_{out}}{P_{in}} = \frac{334.88}{\gamma QH} = \frac{334.88}{9.81 \times 0.8 \times 45}$$

$= 94.8\%$

To find α_2:

$$\tan(180 - \beta_2) = \frac{V_2 \sin \alpha_2}{U - V_2 \cos \alpha_2}$$

$$\tan 15 = \frac{V_2 \sin \alpha_2}{14 + 0.7}$$

$V_{f2} = V_2 \sin \alpha_2 = 3.94$

$$\therefore \tan \alpha_2 = \frac{V_2 \sin \alpha_2}{V_2 \cos \alpha_2} = \frac{3.94}{-0.7}$$

$\alpha_2 \approx -80°$

$or\ \ \alpha_2 = 180 - 80 = 100°$

This means that the velocity triangle at the outlet is as follows,

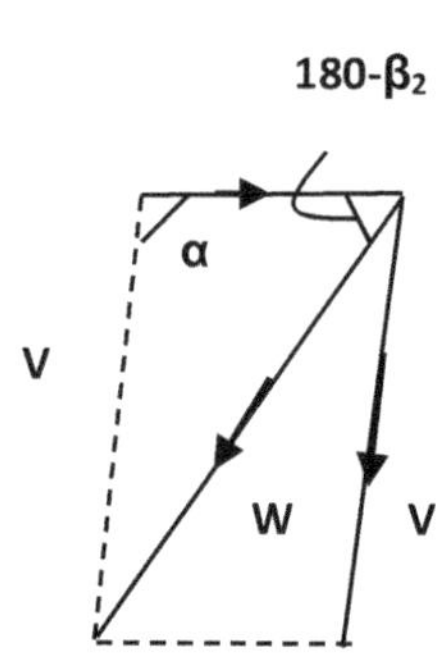

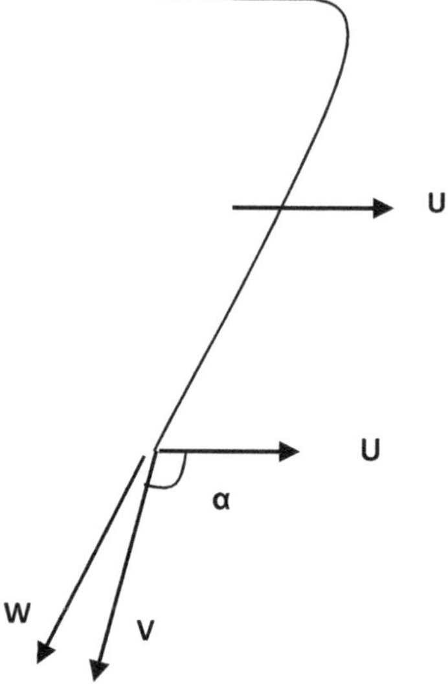

Another solution:

$$F = \rho Q(\overrightarrow{w_1} - \overrightarrow{w_2})$$

$$P_t = F.U = 1000 \times 0.8(w_1 \cos\beta_1 - w_2 \cos\beta_2)U$$

$$w_1 = w_2 = V_1 - U = 29.2 - 14 = 15.2\ m/s$$

$$\therefore P_t = 1000 \times 0.8 \times 15.2\big(1 - (-0.966)\big)14$$

$$= 334.88\ kW$$

$$\eta_t = \frac{P_{out}}{P_{in}} = 94.8\%$$

Example 2: A single jet Pelton turbine is required to drive a generator to develop 10000 kW. The available head at the nozzle is 760 m. Assuming that the efficiency of the electric generator is 95%, Pelton wheel efficiency is 87%, coefficient of velocity for the nozzle is 0.97, mean bucket velocity is 0.46 of the jet velocity, outlet angle of the bucket is 165° and the relative velocity of the water leaving the bucket is 0.85 of that at the inlet, find:

(a) The diameter of the jet.

(b) The force exerted by the jet on the bucket.

If the ratio of the mean circle diameter of the bucket to the jet diameter is not less than 10, find the speed?

Solution:

$P_g = 10000\ kW;\ H = 760\ m;\ \eta_g = 0.95$

$C_V = 0.97;\ \emptyset = 0.46;\ \eta_t = 0.87$

$\beta_1 = 0;\ \alpha_1 = 0;\ w_2 = 0.85w_1;\ m = 10$

The output of the turbine $P_t = \dfrac{P_g}{\eta_g} = \dfrac{10000}{0.95} = 10526\ kW$

$$\eta_t = \frac{P_{out}}{P_{in}} = \frac{10526}{9.81 \times Q \times 760} = 0.87$$

$Q = 1.62\ m^3/s$

$$\therefore d_i = \sqrt{\frac{4Q}{\pi \times C_V\sqrt{2gH}}} = \sqrt{\frac{4 \times 1.62}{\pi \times 0.97\sqrt{2 \times 9.81 \times 760}}}$$

$$= 0.132\ m$$

The force exerted by the jet is given by,

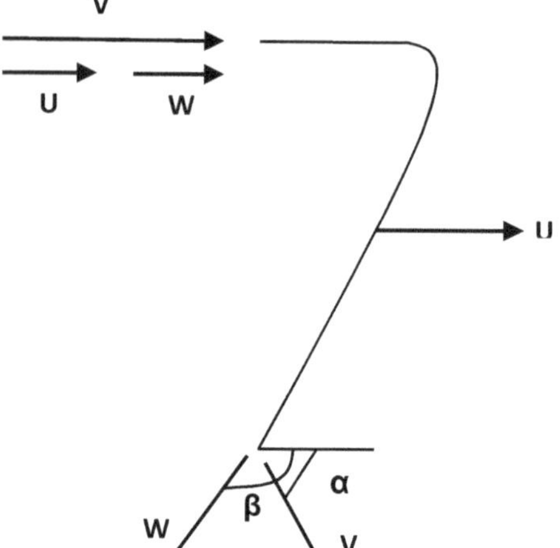

$$F = \rho Q(V_{u1} - V_{u2})$$

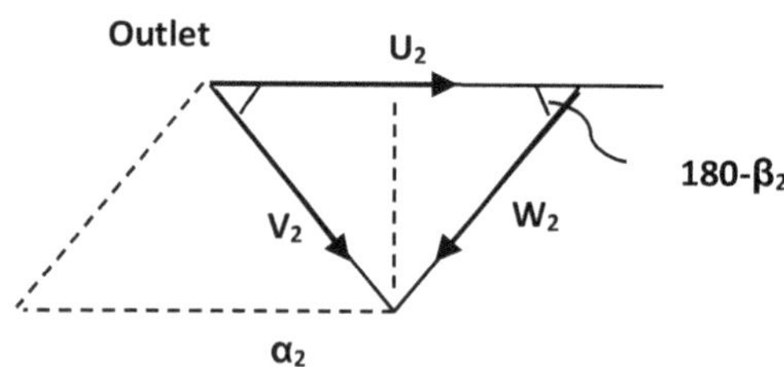

$$U_1 = U_2 = U$$
$$w_1 = w_2$$

Outlet
U_2
180-β_2
V_2
W_2
α_2

$$V_1 = C_V\sqrt{2gH} = 0.97\sqrt{2 \times 9.81 \times 760} = 118.5\ m/s$$

$$U_1 = \emptyset\sqrt{2gH} = 0.46\sqrt{2 \times 9.81 \times 760} = 54.5\ m/s$$

From the velocity triangle at the inlet $\alpha_1 = 0 \quad \beta_1 = 0$

$$w_1 = V_1 - U = 118.5 - 54.5 = 64\ m/s$$

$$w_2 = 0.85 w_1 = 0.85 \times 64 = 54.6\ m/s$$

$and \quad V_{u1} = V_1 \cos\alpha_1 = 118.5\ m/s$ From the outlet velocity triangle :

$$V_{u2} = U_2 - w_2 \cos(180 - \beta_2)$$

$$= 54.5 - 54.6\,cos15$$

$$= 1.83\ m/s$$

$$\therefore F = 1000 \times 1.629 \times (118.5 - 1.83)$$

$$= 189\ kN$$

$$m = \frac{D_1}{d_1} = 10 \qquad \therefore D_1 = 1.32\ m$$

$$U = \frac{\pi D_1 N}{60} \qquad or\ \ N = \frac{60 \times 54.5}{\pi \times 1.32}$$

$$N = 787\ rpm$$

2.3. POWER REGULATION MECHANISMS

The flow rate through the Pelton turbine and the power of the turbine are regulated by shifting the needles in the permitted direction, by which the flow area of the nozzles is changed. Formerly, use was made, as a rule of the needle operating gear shown schematically in Fig. (**2.5**). Needle 2 is fixed at the end of the long rod 3, which is connected to the piston of hydraulic servomotor 4. The needle is aligned and its radial displacement is prevented by a guide sleeve and spacer ribe 5. This design requires a rather sharp turn of the water supply conduit directly upstream of nozzle 1, and this is due to the origination of reverse flows at the turn, diminishes the density of the jet at the nozzle outlet and impairs the power-generating characteristics of the turbine [3].

In the last few decades, use has been made of the so-called straight flow nozzle 1, as shown in Fig (**2.5a**). Needle 2 has fasted on the short rod 3 connected to the piston of servomotor 4. The servomotor is made in the form of a bulb secured in the center of a tube by ribs 5. The bulb is exposed to water flow and this makes it unnecessary to turn water supply conduit, as shown in Fig (**2.6**).

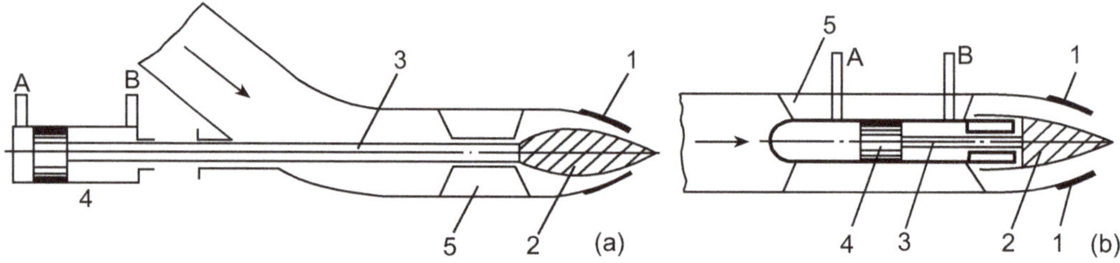

Fig. (2.5). Schematic diagram of Pelton turbine regulating needle gear.

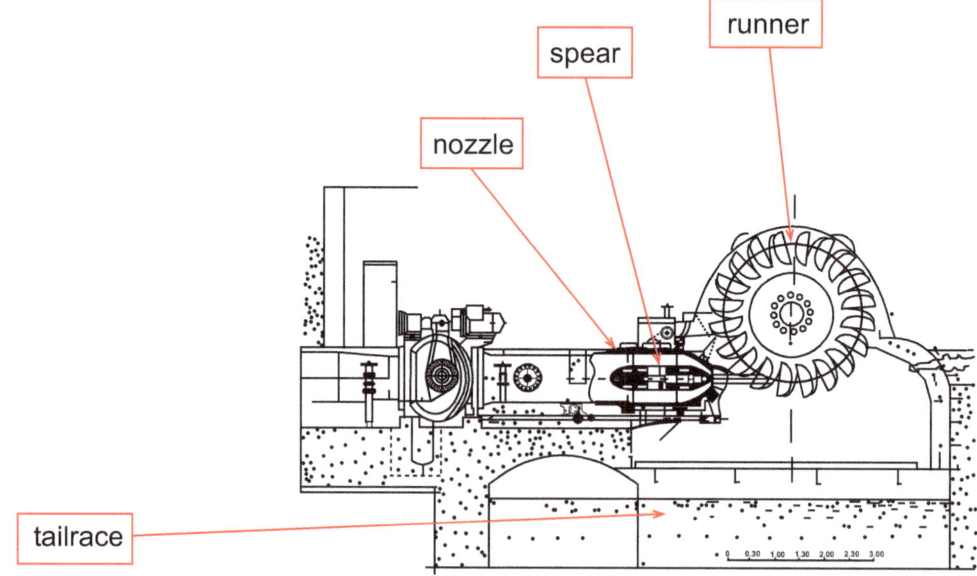

Fig. (2.6). Horizontal axis 1-jet turbine [3].

2.4. SUPPLY AND DISCHARGE SYSTEM

The water is brought in the penstock ending in a single or multi-nozzles. The total pressure energy of the water flowing from the nozzle produces a free water jet. The free water jet is, in turn, made to strike on a series of buckets mounted on the periphery of the turbine wheel. Therefore, the casing of an impulse turbine has no hydraulic function to perform. It is necessary only to prevent water splashing, lead the water to the tailrace, and act as a safeguard against the accident of the shape and size of the turbine case and the penstock must be fitted with the layout of the house of the HEPP, (Fig. **2.7**).

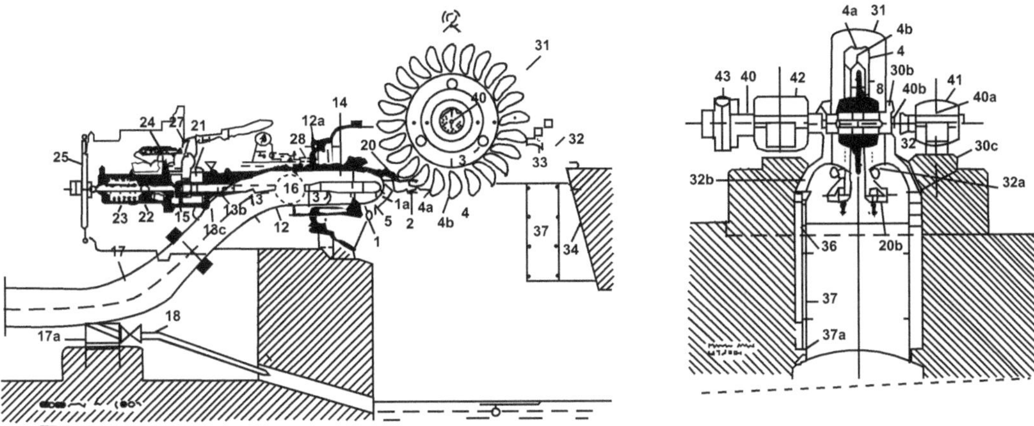

Fig. (2.7). The horizontal arrangement of the Pelton Turbine [4]. 1 Nozzle1a, Inset pice 1b Holder 1c Eyes 1d Protecting roof 2 Water jet 3 Runner wheel 4 Runner buckets 4a Shaped-out portion of bucket 4b Center partition 5 spearhead 5a spear Point 5c Stud bolt 5d Key 5e Spear body 5f,5g Cylindrical holes with pins 6 Tail race D_1 Runner diameter d. jet diameter 7 Lugs of bucket 8 Runner disc 8a Runner boss 9 Fixing bolts with nuts 9a and 9b Bolt head 10 Tensioning bolts 11 Radial keys 12 Inlet bend 12a Flange 13 Spear rod 13a Bronze sleeve 13b Balancing piston 13c Leather packing 14 Guide cross 14a Guide ribs 14b Cylindrical member for foregoing 15 Support for spear rod 16 Cleaning hole 17 Lower bend 17a Anchoring foot 18 Emptying pipe 19 Turbine valve 20 Jet deflector 20a Intercepting piece 20b Levere 21 Relay lever 22 Servomotor piston for spear adjustment 23 Closing spring for spear 24 Distributing valve for nozzle 25 Handwheele for spear adjustment 26 Deflector rod 27 Cam 30 Casing 30a Opening for the inlet bend 30b Lateral chambers 30c Cast iron bearing supports 31 Casing hood 32 Lower part of casing 32a Guide walls 32b Drainage channels for spray water from turbine shaft 33 Spray water catcher 34 Lining 36 Cooling coils 37 Cover for protecting of latter 37a Hole for discharging cooling water 40 Turbine shaft collar 40b Ring 41 External turbine bearing 42 Internal turbine bearing 43 Shaft coupling 44 Speed governor 45 Flywheel.

The horizontal arrangement is found only in the medium and small size turbines with usually one or two jets. In some designs, up to four jets are used. The flow passes through the inlet bend to the nozzle outlet, where it flows out as a compact through atmospheric air on to the heel buckets. From the outlet of the buckets, the water falls through the pit down into the tail water canal.

Part of the Pelton turbine and different types of jet distribution are shown in Figs. (**2.8-2.11**).

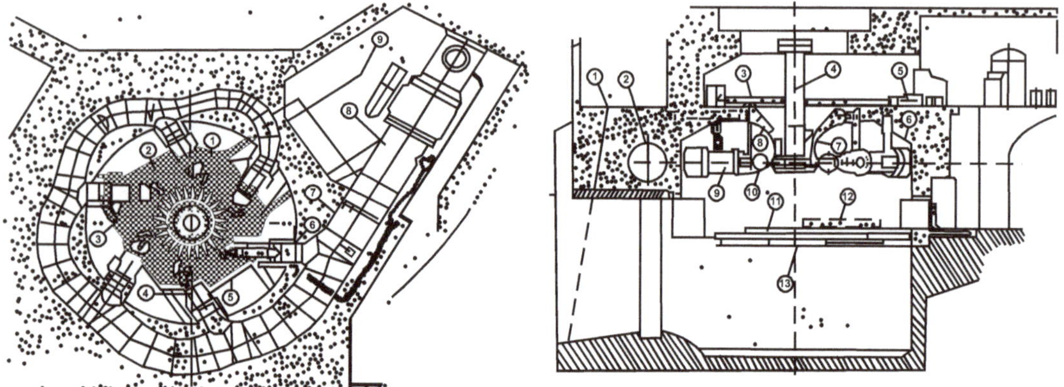

Fig. (2.8). The vertical arrangement of the Pelton turbine [4].

Large Pelton turbines with many jets are normally arranged with a vertical shaft. The jets are symmetrically distributed around the runner to balance the jet forces. Fig. (**2.7**) shows the vertical and horizontal sections of the arrangement of a six jets vertical Pelton turbine.

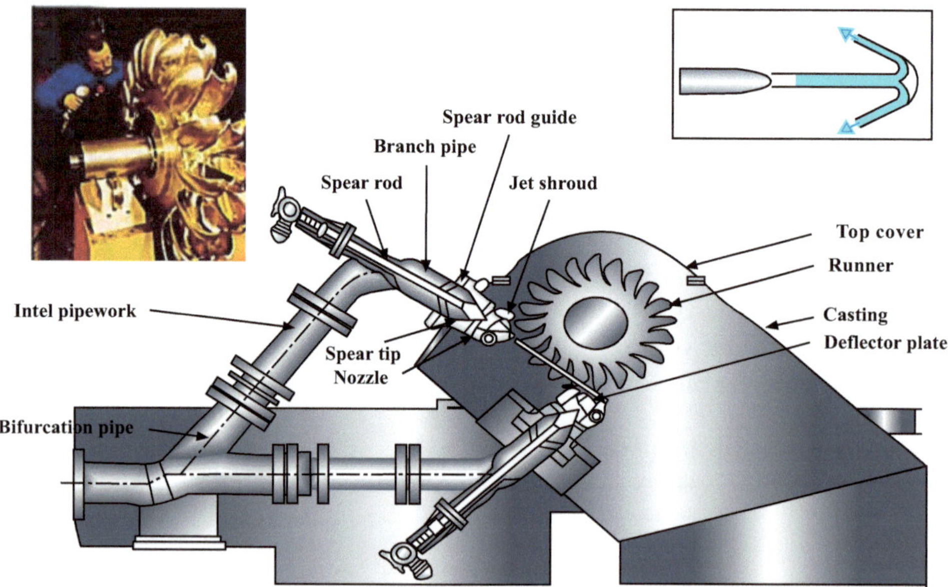

Fig. (2.9). Parts of the Pelton Turbine [5].

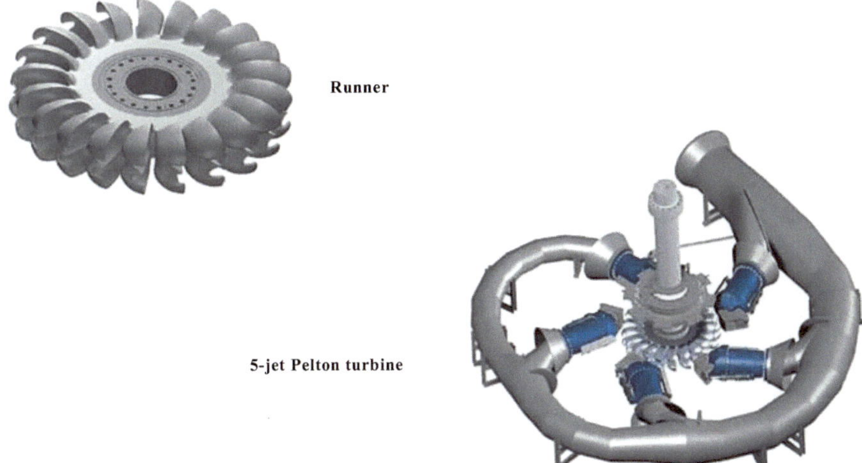

Source: Volth-Siomons

Fig. (2.10). Pelton Turbine Components [5].

Fig. (2.11). Pelton Turbine 5-jet turbine [5].

SOLVED PROBLEMS

Q.1: A power hose is equipped with impulse turbine of the Pelton type. Each turbine delivers a maximum power of 14.2 MW when working under a head of 855 m and running at 600 rpm. Find the least diameter of the jet and the mean diameter of the wheel. What would be the approximate diameter of the orifice of the nozzle tip? Determine the value of the jet ratio and state if it is within the limits. Specify the number of buckets for the wheel. Take the overall efficiency of the turbine as 89.2%, coefficient of speed ratio as Φ=0.45, and velocity coefficient as $C_V = 0.988$.

Solution:-

$P_t = 14.2\ MW; H = 855\ m$

$N = 600\ rpm\ ; \eta_t = 0.892$

a) Least diameter of the jet d_1:

$P_t = \gamma Q H \eta_t$

$14200 = 9.81 \times 855 \times 0.89 \times Q$

$Q = 1.895\ m^3/s$

Further;

$$Q = a_1 V_1 = \frac{\pi}{4} d_1^{\ 2} . C_V \sqrt{2gH}$$

$$1.895 = \frac{\pi}{4} d_1^{\ 2} \times 0.988\sqrt{2 \times 9.81 \times 855}$$

$d_1 = 0.1372\ m = 137.2\ mm$

b) Mean diameter of the wheel D_1

$$U_1 = \emptyset\sqrt{2gH} = \frac{\pi D_1 N}{60}$$

$$D_1 = \frac{60 \times 0.45 \times \sqrt{2 \times 9.81 \times 855}}{\pi \times 600}$$

$$D_1 = 1.85\ m \quad or\ 1850\ mm$$

c) Diameter of the nozzle $d_0 = 1.25 d_1 = 1.25 \times 13.72$

$$= 17.2\ cm$$

d) Jet ratio $m = \frac{D_1}{d_1} = \frac{1.85}{0.1372} = 13.5$

e) $Z = 15 + 0.5\ m = 21.75 \approx 22\ buckets;$ $\quad Z = No.of\ \ buckets$

Q.2: A double overhung Pelton turbine unit is used to operate a 30 MW generator under an effective head of 300 m at the base of the nozzle. Find the size of the jet, mean diameter of runner, synchronous speed, and specific speed of each wheel. Assume the generator efficiency is 93%, the Pelton wheel efficiency is 85%, the coefficient of nozzle velocity is 0.97, speed ratio is 0.46 and jet ratio is 12.7.

Solution:

There are two runners keyed on the two ends of the shaft, and the generator lies between them. Each runner is to be taken as one complete turbine. Thus, the generator is fed by two Pelton turbines

∴ The developed power by each turbine is,

$$P_t = \frac{30000}{2 \times 0.93} = 16130\ kW$$

and the available power of each turbine is,

$$P_a = \frac{16130}{0.85} = 18975\ kW$$

$$= \gamma Q H_n \qquad \therefore Q = \frac{18975}{9.81 \times 30}$$

$$Q = 6.47\ m^3/s$$

The velocity of the jets are expressed as,

$$V_1 = C_V\sqrt{2gH} = 0.97\sqrt{2 \times 9.81 \times 300}$$

$$V_1 = 74.3\ m/s$$

$$and\ \ Q = a_1 V_1 = \frac{\pi}{4} {d_1}^2 \times 74.3$$

$$\therefore\ d_1 = 0.333\ m$$

For the jet ratio,

$$m = \frac{D_1}{d_1} = 12 \quad \therefore\ D_1 = 12 \times 0.333 = 4\ m$$

$\therefore U_1 = \emptyset\sqrt{2gH} = \frac{\pi D_1 N}{60}$ is the peripheral velocity of the wheel

$$\therefore U_1 = 0.46 \times \sqrt{2 \times 9.81 \times 300} = \frac{\pi D_1 N}{60} = 35.3$$

so that,

$$N = \frac{60 \times 35.3}{\pi \times 4} = 168.5\ rpm$$

$$f = \frac{P.N}{60}\ \ where\ P\ No. of\ Poles = 18\ (assume)$$

$$\therefore N_{syn} = \frac{60 \times f}{P} = \frac{3000}{18} = 166.7\ rpm\ where\ f = 50\ Hz$$

Revised Diameter of the wheel

$$D_1 = \frac{168.5 \times 4}{166.7} = 4.05\ m$$

Specific speed $N_s = \frac{N\sqrt{P_t}}{H^{\frac{5}{4}}} = \frac{166.7\sqrt{16130}}{(300)^{\frac{5}{4}}}$

$$N_s = 17\ m - kWunit$$

Q.3: A Pelton wheel is to be designed to generate 750 kW at 400 rpm. It is to be supplied with water from a reservoir whose level is 250 m above the wheel through a pipe of 900 m long. The pipeline losses are to be 5% of the gross head. The coefficient of friction is 0.02. The bucket speed is 0.46 of the jet speed, and the efficiency of the wheel is 85%. Calculate the pipeline diameter, jet diameter, and wheel diameter, C_V=0.985.

Solution:-

$$P_t = 750\ kW\ \ ; N = 400\ rpm\ \ ; H_{gross} = 250\ m$$

$$L = 900\ m\ \ ;\ h_f = 5\%\ of\ gross\ head\ ; f = 0.02$$

$$\emptyset = 0.46\ ;\ \eta_t = 0.85\ ;\ C_V = 0.985$$

$$a)\ Pipeline\ losses\text{:}\ \ h_f = f.\frac{L}{D}.\frac{V^2}{2g} = \frac{8fLQ^2}{\pi^2 gD^5} = 0.05 \times 250 = 12.5\ m$$

$$but\ \ Q = \frac{P_t}{\gamma H\eta_t} = \frac{750}{9.81(0.95 \times 250) \times 0.85}$$

$$= 0.379\ m^3/s$$

$$\therefore 12.5 = \frac{8 \times 0.02 \times 900 \times (0.379)^2}{\pi^2 \times 9.81 \times D^5}$$

$$\therefore D = 0.44\ \text{m}\ \ \ or\ 440\ mm$$

b) Jet ratio: $d_1 = \sqrt{\frac{Q}{\frac{\pi}{4}(C_V\sqrt{2gH})}}$

$$d_1 = \sqrt{\frac{0.379}{\frac{\pi}{4}0.985\sqrt{2 \times 9.81 \times 0.95 \times 250}}} = 0.0847\ m\ = 84.7\ mm$$

c) Wheel diameter D_1:

$$U_1 = \emptyset\sqrt{2gH}$$

$$= 0.46\sqrt{2 \times 9.81 \times 237.5}$$

$$= 31.4\ m/s$$

$$and\ \ U_1 = \frac{\pi D_1 N}{60}$$

$$\therefore\ D_1 = \frac{60 \times 31.4}{\pi \times 400} = 1.5\ m$$

Q.4: Determine the discharge, least jet diameter, mean runner diameter, jet ratio, and number of buckets for the following Pelton turbine:

Power =127 kW, Head =300 m, Speed =600 rpm

Assume the following constants:

C_V=0.98, Φ =0.45, η_t=0.75.

Solution:-

$$P_t = \gamma Q H \eta_t$$

a) Discharge: $$Q = \frac{P_t}{\gamma H \eta_t} = \frac{127}{9.81 \times 300 \times 0.75} = 0.0575\ m^3/s$$

b) Least jet diameter:

$$Q = \frac{\pi}{4}d_1{}^2.\left(C_V\sqrt{2gH}\right)\ where\ V_1 = \left(C_V\sqrt{2gH}\right)$$

$$\therefore d_1 = \sqrt{\frac{0.0575}{\frac{\pi}{4} \times 0.98 \times \sqrt{2 \times 9.81 \times 300}}} = 0.0312\ m\ = 31.2\ mm$$

c) Mean runner diameter:

$$U_1 = \frac{\pi D_1 N}{60} = \emptyset\sqrt{2gH}$$

$$\therefore \frac{\pi D_1 \times 600}{60} = 0.45\sqrt{2 \times 9.81 \times 300}$$

$$D_1 = 1.1\ m$$

d) Jet ratio:

$$m = \frac{D_1}{d_1} = \frac{1.1}{0.0312} = 35.3$$

e) Bucket Number:

$$Z = 0.5 \times m + 15 = 0.5 \times 35.3 + 15 = 33$$

Q.5: Show that in Pelton wheel, where the bucket deflect the water through an angle of (180^0- β_2), the hydraulic efficiency of the wheel is given by

$$\eta_h = \frac{2U(V-U)(1+\cos\theta)}{V^2}$$

Where, V is the velocity of jet and U is the velocity of the wheel at the pitch radius. If the bucket speed is 30 m/s, β_2=160° and the jet diameter is 12.5 cm. Find the jet velocity for maximum efficiency and the corresponding water power in kW of the turbine.

Solution:-

a) Proof: $\alpha_1 = 0;\ \beta_1 = 0$

For Pelton turbine:

$U_1 = U_2$
$W_1 = W_2$

$and\ \ W_1 = V_1 - U_1$

$V_{u1} = V_1 cos\alpha_1 = V_1$

Velocity diagram:

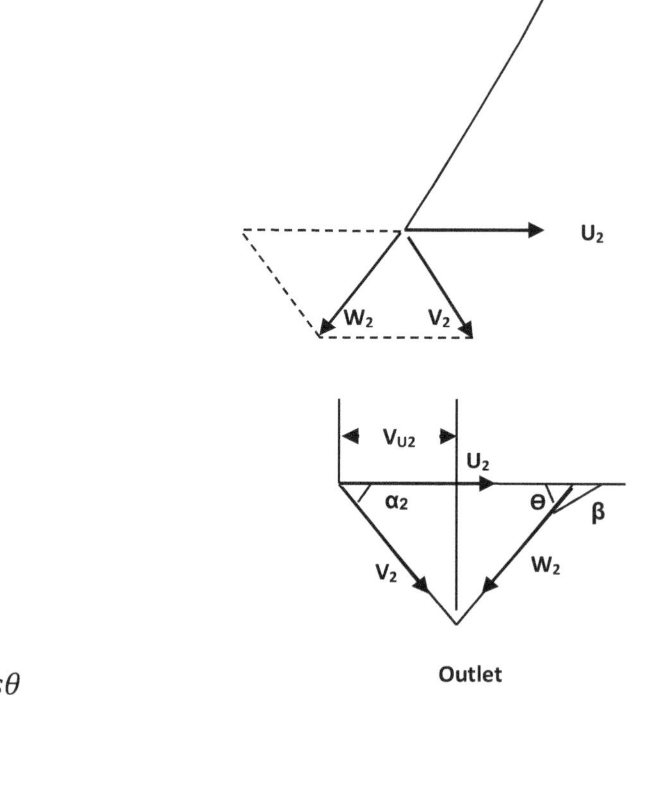

$V_{u2} = V_2 cos\alpha = U_2 - W_2 cos\theta$

$= U_1 - (V_1 - U_1)cos\theta$

Hydraulic power:

$$P_h = \rho Q(V_{u1} - V_{u2})U_1$$

$$= \rho Q(V_1 - (U_1 - (V_1 - U_1)cos\theta)U_1$$

$$= \rho Q((V_1 - U_1) + (V_1 - U_1)cos\theta)U_1$$

$$P_h = \rho Q(V_1 - U_1)(1 + cos\theta)U_1 \ \ldots\ldots\ldots (1)$$

Available power $P_a = \gamma QH = \rho gQ.\dfrac{V_1^{\ 2}}{2g} = \rho Q\dfrac{V_1^{\ 2}}{2} \ \ldots\ldots\ldots (2)$

$$\eta_h = \frac{P_h}{P_a} = \frac{\rho Q(V_1 - U_1)(1 + cos\theta)U_1}{\rho Q \frac{V_1^{\ 2}}{2}}$$

$$\eta_h = \frac{2U(V - U)(1 + cos\theta)}{V^2}$$

b) for maximum efficiency:

$$\frac{d\eta_h}{du} = 0 \quad \therefore \frac{V_1}{U_1} = 2$$

$$\therefore V_1 = 2U_1 = 2 \times 30 = 60\, m/s$$

c) Available power: $P_a = \gamma QH = \gamma Q.\frac{V_1^{\ 2}}{2g}$

$$Q = \frac{\pi}{4} d_1^{\ 2} V_1^{\ 2} = \frac{\pi}{4}(0.125)^2 \times 60 = 0.736\, m^3/s$$

$$\therefore P_a = \rho Q \frac{V_1^{\ 2}}{2} = 1 \times 0.736 \times \frac{60^2}{2} = 1325\, kW$$

Q.6: A Pelton wheel supplied with water in the nozzle box at a pressure head of 150 m. The velocity coefficient for the nozzle is 0.96. The relative velocity of water while in contact with the bucket is turned through 150° and is also reduced by 15% due to the friction losses. The required ratio of the bucket to jet-speed is 0.47 and 0.8 of the energy imparted to the wheel as determined by the velocity diagram which is available at the outlet shaft. The bucket circle diameter is ten times the jet diameter. Find the jet and wheel diameters, and the speed of rotation to develop 150 kW under these conditions.

Solution:-

$$H = 150\, m\ ;\ C_V = 0.96\ ;\ \beta_2 = 150°\ ;\ w_2 = 0.85w_1$$

$$\frac{U}{V} = 0.47 \quad or \quad \emptyset = 0.96 \times 0.47 = 0.45$$

$$\eta_{mech} = 0.8 \ ; \ m = \frac{D_1}{d_1} = 10 \ ; \ P_t = 150\ kW$$

a) jet dia d_1 : $Q = a_1 V_1 = \frac{\pi}{4} {d_1}^2 V_1$

$$V_1 = C_V \sqrt{2gH} = 0.96\sqrt{2 \times 9.81 \times 150} = 52.1\ m/s$$

$$\therefore U_1 = 0.47 \times 52.1 = 24.5\ m/s = U_2$$

For velocity diagram

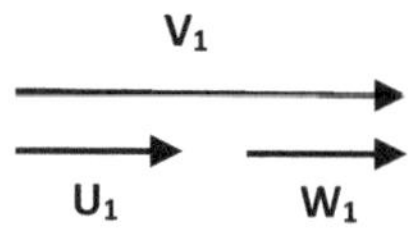

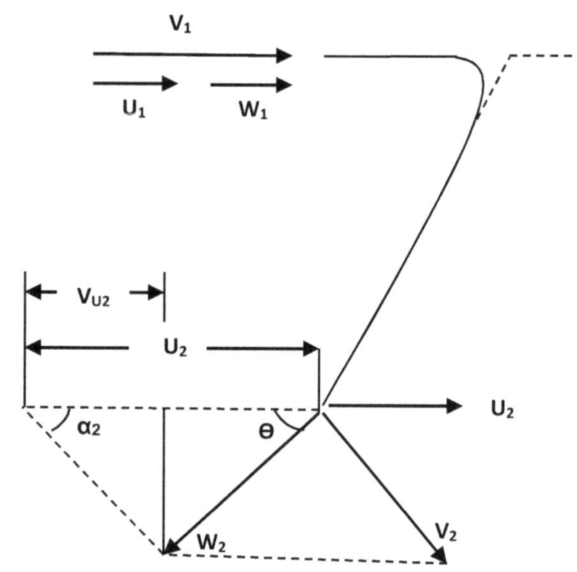

$V_{u1} = V_1 cos\alpha_1 = V_1$

$\alpha_1 = 0$

$w_1 = V_1 - U_1 = 27.6\ m/s$

$V_{u2} = U_2 - W_2 cos\theta$

$w_2 = 0.85 w_1$

$= 0.85(V_1 - U_1)$

$= 0.85(52.1 - 24.6)$

$= 23.46\ m/s$

$\therefore V_{u2} = 24.5 - 23.46 \cos 30 = 4.2\ m/s$

Hydraulic power: $P_h = \rho Q (V_{u1} - V_{u2}) U_1$

$$= 1000 \times Q(52.1 - 4.2)24.5$$

$$= 1173.55\ Q\ kW$$

$$since\ \eta_{mech} = 0.8 \quad \therefore P_t = 0.8\ P_h = 938.84\ Q\ kW$$

$$or\ \ 150 = 938.34\ Q$$

$$or\ Q = 0.16\ m^3/s$$

$$\therefore\ d_1 = \sqrt{\frac{0.16}{\frac{\pi}{4}(52.1)}} = 0.062\ m = 62\ mm$$

b) $D_1 = 10\ d_1 = 620\ mm$

c) $U_1 = \frac{\pi D_1 N}{60} = 24.5$ m/s

$$\therefore N = \frac{60 \times 24.5}{\pi \times 0.62}$$

$$N = 755\ rpm$$

Q.7: A Pelton turbine has a mean bucket speed of 40 m/s and is supplied with water at the rate of 0.55 m^3/s, the head of water behind the nozzle is 400 m. If the jet is deflected by the head of water through 170°, find the power developed and the efficiency of the wheel. Assume β_1=0, C_V=0.985.

Solution:-

$$U_1 = U_2 = 40\ m/s \quad ;Q = 0.55\ m^3/s$$

$$H = 400\ m\ ;\ \beta_2 = 170°\ ;C_V = 0.985$$

$$V_1 = C_V\sqrt{2gH} = 0.985\sqrt{2 \times 9.81 \times 400} = 87.27\ m/s$$

$$and\ \ U_1 = U_2 \quad ; \qquad w_1 = w_2$$

$V_{u1} = V_1 cos\alpha_1 = V_1 \qquad \alpha_1 = 0$

and from velocity triangle

$V_{u2} = U_2 - W_2 \cos\theta$

$= 40 - (87.27 - 40)\cos 10$

$= -6.55\ m/s$

Power developed by the wheel $P_h = \rho Q(V_{u1} - V_{u2})U_1$

$$= 1000 \times 0.55(87.27 + 6.55) \times 40$$

$$= 2064\ kW$$

$$\text{Available power } P_a = \gamma QH = 9.81 \times 0.55 \times 400 = 2158\ kW$$

$$\eta_h = \frac{P_h}{P_a} = \frac{2064}{2158} = 95.6\%$$

Q.8: A Pelton turbine has a mean speed of 12.2 m/s and supplied with water at the rate of 1370 l/s under a head of 30.5 m. If the bucket deflects the jet through an angle of 160°, f, find the outlet power and the efficiency of the wheel. C_V=0.985, α_1=0.

Note that the number of buckets are Z=0.5 m+15

Solution:-

$$U_1 = U_2 = 12.2\ m/s \quad ; Q = 1.37\ m^3/s$$

$$H = 30.5\ m\ ;\ \beta_2 = 160°\ ; \alpha_1 = 0\ ;\ C_V = 0.985$$

$$W_1 = W_2 = V_1 - U_1$$

$$and\ V_1 = 0.985\sqrt{2 \times 9.81 \times 30.5}$$

$V_1 = 24.1\ m/s \quad and\ W_1 = 11.9\ m/s$

$also\ V_{u1} = V_1 cos\alpha_1 = 24.1$ m/s

$V_{u2} = U_2 - W_2 \cos\theta$

$= 12.2 - 11.9 \cos 20$

$= 1\ m/s$

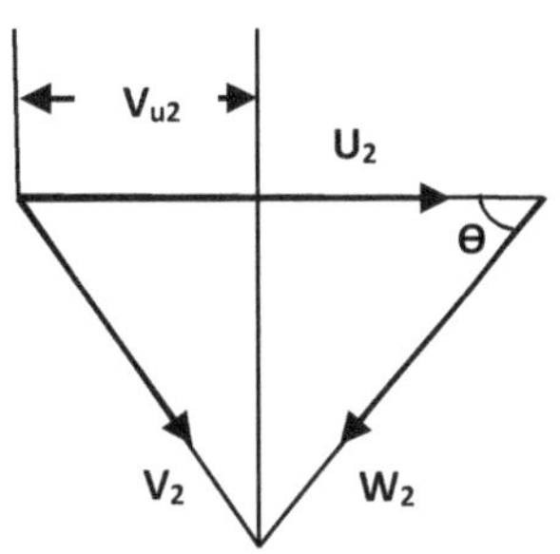

$$\therefore \text{Available power } P_a = \gamma QH$$

$$= 9.81 \times 1.37 \times 30.5 = 410\ kW$$

$$\eta_h = \frac{386}{410} = 94.1\%$$

CHAPTER 3

Reaction Turbines

Abstract: In a reaction turbine, the runner utilizes both potential and kinetic energies. As flows through the stationary part of the turbine, the whole of its pressure energy is not transformed into kinetic energy and when the water flows through the moving parts, there is a change both in pressure and in the direction and velocity of flow of water. As the water gives up its energy to the runner, both its pressure and absolute velocity were reduced. The water, which acts on the runner blades is under a pressure above atmospheric and the runner passages are always completely filled with water.

The important reaction turbines are Francis and Kaplan which are discussed in this chapter according to their specification related to hydro–electric power plant. Theory for each type presented with sort notes and solved problems.

Keywords: Draft Tube, Flowrate through Reaction Turbine, Net Head, Reaction Turbine, Supply and discharge systems, Velocity Triangle.

3.1. TYPE OF REACTION TURBINE

In general, there are two types of reaction turbines, Francis and Kaplan turbines according to the direction of flow. The water enters the runner under pressure and flows over the vanes, the pressure head of water while flowing over the vanes, is converted into velocity head and finally reduced to the atmospheric pressure.

1. Francis Turbine: - it is an inward flow reaction turbine having radial discharge at the outlet. It is operating under a medium head and required medium quantity of water

Flow rate $\approx 2 \rightarrow 800\ m^3/s$

Head $\approx 50 \rightarrow 400$ m

Net power up to ≈ 800 MW

2. Kaplan Turbine: - it is an axial flow reaction turbine in which the flow of water is parallel to shaft. A Kaplan turbine is used where a large quantity of water is available at a low head.

Jafar Mehdi Hassan, Salman Hussien Omran, Laith Jaafer Habeeb,
Alamaslamani Ammar Fadhil Shnawa & Adrian Ciocănea

Flow rate up to ≈ 1000 m^3/s

Head ≈ 40 m

Net power up to ≈ 200 MW

All parts of the Kaplan turbine such as spiral casing, guide mechanism and draft tube are similar as in Francis turbine system except:

The runner: the runner has two major differences. In Francis runner, the water enters radially while in Kaplan type it strikes the blades axially. The number of vanes in Francis turbines is 16 → 24 while in Kaplan it is only 3 → 6 vanes or at most 8 in an exceptional case. Thee RPM more than twice than that of Francis turbine.

Kaplan turbines have taken the place of Francis turbine for certain medium head installation [3].

3.2. CONSTRUCTION OF REACTION TURBINE

The turbine systems have the following component (Figs. **3.1** and **3.2**): -

1. **Penstock**: it is a waterway to carry water from the reservoir to the turbine casing. The Penstock section was manufactured in quarters and welded at the site (*e.g*. the penstock of Hydropower station in Venezuela is 7.5 m in diameters).

2. **Spiral Casing or Scroll Casing**: - to avoid losses of efficiency, the scroll casing is designed with a cross-sectional area reducing uniformly around the circumference, maximum at the entrance and nearly zero at the tip. This gives a spiral shape and hence the casing is named spiral casing.

3. **Guide Mechanisms**: - the Guide vanes or wicket gates are fixed between two rings known as guide wheel. The guide vane can be closed or opened to allowing a variable quantity of water according to the needs. The guide mechanism parts are: - Guide vanes, Guide wheel, regulating shaft, and Governor.

4. **Runner and Turbine Main Shaft**: - the aim of the runner is to guide the flow

inside the turbine. The width of the runner depends upon the specific speed. The runner may be classified as (i) slow (ii) medium (iii) fast, depending upon the specific speed. The runner is keyed to the shaft which may be vertical or horizontal.

5. **Draft Tube**: - the water after passing through the runner flows down through a tube called draft tube. It is generally drowned a proximately 1 m below the tail rase level. The advantage of it:

    a) It increases the head of water by an amount equal to the height of the runner outlet above the tail race.

    b) It increases the efficiency of the turbine [4].

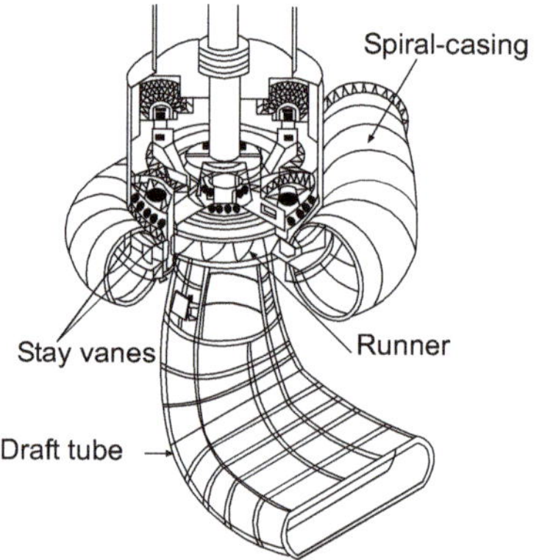

Fig. (3.1). Major Components of a Francis Turbine.

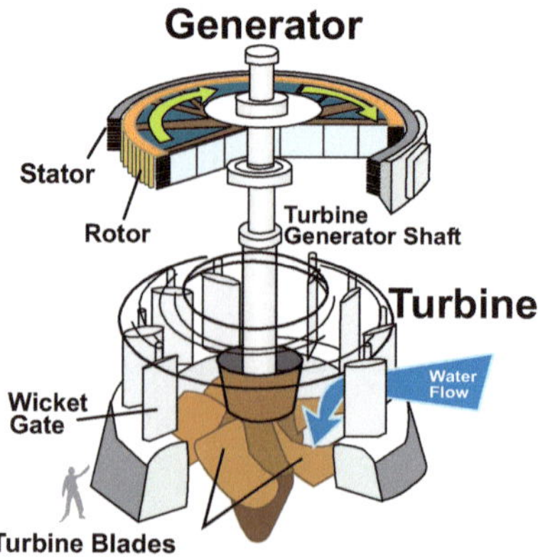

Fig. (3.2). Diagram of typical Kaplan Turbine.

3.3. THEORY OF REACTION TURBINE

The one – dimensional from of steady flow momentum equation is:

$$\sum F = \dot{m}(\vec{V}_{out} - \vec{V}_{in}) \quad \textbf{(3.1)}$$

And for the turbine the force exerted by flowing fluid on the boundary of flow passage due to change of momentum. (reaction force according to Newton 3rd law) then the resultant dynamic force exerted by the water on the runner vanes in the direction of rotation is:

$$-F = R$$

$$\therefore R = \rho Q(\vec{V}_{in} - \vec{V}_{out}) \quad \textbf{(3.2)}$$

Or in general written as

$$or \;\; F = \rho Q(V_{u1} - V_{u2}) \quad \textbf{(3.3)}$$

Where $V_u = V \cos\alpha$ the whirl velocity of flow.

and the hydraulic power developed by the turbine:

$$\therefore \;\; Power \;\; P_t = \rho Q(V_{u1}U_1 - V_{u2}U_2) \quad \textbf{(3.4)}$$

Where $U_1 \neq U_2$ $\quad U = wr \quad w = \frac{\pi DN}{60}$

The power developed by the turbine or may be named hydraulic power considering the loss of head including that due to outlet velocity then:

For a given rate of flow the minimum value of V_2 occurs when V_2 perpendicular to the tangential velocity U_2

$$i.e. \;\; V_{u2} = 0 \quad where \;\; \alpha_2 = 90°$$

$$\therefore P_t = \rho Q V_{u1} U_1 = P_H \;\; hydraulic\ power. \quad \textbf{(3.4a)}$$

$$the\ Work\ done/sec\ per\ \ kN = \frac{V_{u1}U_1}{g}$$

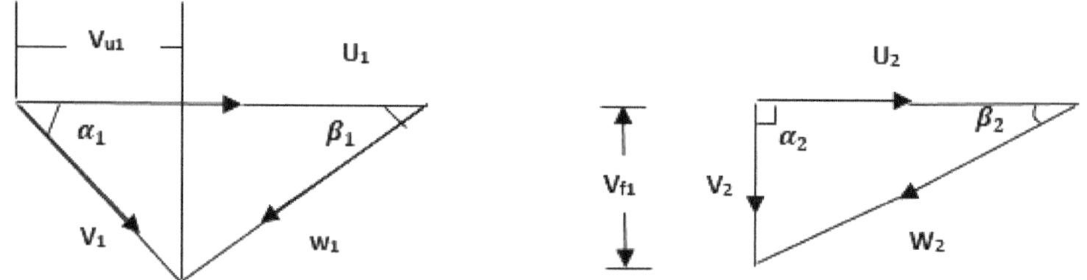

Inlet velocity triangle Outlet velocity triangle

3.4. EFFICIENCY OF REACTION TURBINES

1. Head Efficiency

$$\eta_H = \frac{H - \Delta H}{H} \quad (3.5)$$

H: operating head (net head)

ΔH: losses (friction, kinetic head $\left(\frac{V^2}{2g}\right)$) ….. *etc.*

Since

$$P_H = P_a \eta_H = \gamma Q H \eta_H$$

Or

$$\rho Q(V_{u1}U_1 - V_{u2}U_2) = \gamma Q H \eta_H$$

$$\therefore \quad \eta_H = \frac{V_{u1}U_1 - V_{u2}U_2}{gH} \quad (general) \quad (3.6)$$

Or

$$\eta_H = \frac{V_{u1}U_1}{gH} \quad \text{For radial flow at outlet } \alpha_2 = 90° \quad (3.7)$$

2. Hydraulic Efficiency

Total hydraulic loss in turbine made up of total head loss and Q leakage.

i.e.

$$P_h = \gamma(Q - \Delta Q)(H - \Delta H)$$

$$P_a = \gamma Q H$$

$$\therefore \quad \eta_h = \frac{\gamma(Q-\Delta Q)(H-\Delta H)}{\gamma QH} = \eta_Q \eta_H \tag{3.8}$$

3. Volumetric Efficiency

$$\eta_Q = \frac{Q-\Delta Q}{Q} \tag{3.9}$$

ΔQ: volumetric loss due to the clearance.

If the volumetric losses neglected

Then $\eta_H = \eta_h = \frac{V_{u1}U_1}{gH}$ it is the general formula.

4. Mechanical Efficiency

$$\eta_{mech} = \frac{P_h - \Delta P_{mech.loss}}{P_h} \tag{3.10}$$

$$= 1 - \frac{\Delta P_{mech.loss}}{P_h}$$

Where the power developed by the turbine or (brake power)

$$P_t = P_h \eta_{mech} \quad kW$$

5. Overall Efficiency

$$\eta_t = \frac{P_t}{P_a} \tag{3.11}$$

And

$$\eta_t = \eta_Q \eta_h \eta_{mech} \tag{3.12}$$

3.5. FLOW-RATE THROUGH REACTION TURBINE

1. Francis Turbine

$$Q = \pi D B V_f \times contraction \text{ factor} \tag{3.13}$$

$$or \quad Q = \pi D_1 B_1 V_{f1} \times contraction \text{ factor}$$

$$Q = \pi D_2 B_2 V_{f2} \times contraction \text{ factor}$$

Where

$$B = width\ or\ depth\ or\ breadth\ of\ the\ runner.$$

For exact solution

$$Q = (\pi D - Zt)BV_f \tag{3.14}$$

$$Z = number\ of\ vanes\ or\ blades\ in\ the\ runner$$

$$t = thickness\ of\ the\ vanes.$$

$$V_f = V \sin\alpha\ flow\ velocity.$$

2. Kaplan Turbine

The Kaplan turbine is shown in Fig. (**3.3**).

$$Q = \frac{\pi}{4}\left({D_1}^2 - {D_0}^2\right)V_F \tag{3.15}$$

$$D_1 = Outlet\ diameter\ of\ the\ runner.$$

$$D_0 = Diameter\ of\ the\ shaft(Boss)$$

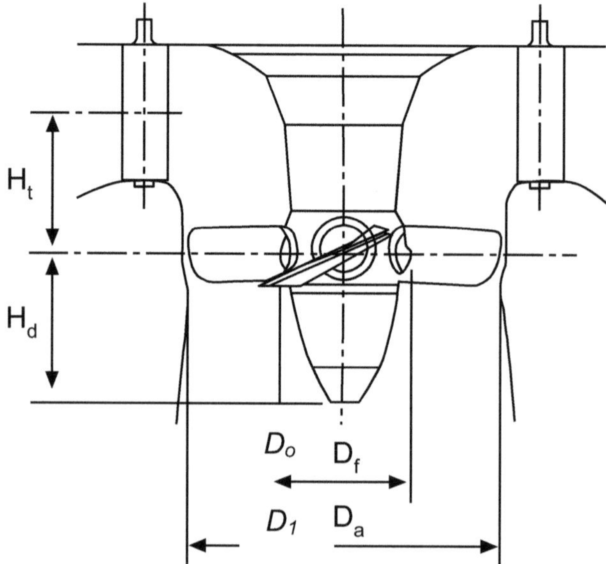

Fig. (3.3). Schematic diagram for Kaplan turbine runner.

3.6. VELOCITY TRIANGLE FOR REACTION TURBINE

The absolute velocity at exit of the runner is such that there is no whirl at the outlet *i.e.* $V_{u2} = 0$; $\alpha_2 = 90°$ then the power

$$P_t = \rho Q(U_1 V_{u1} - U_2 V_{u2}) \tag{3.16}$$

w=w normal as shown in following figures are the rotational speed for which the turbine give the lowest energy loss at outlet represented mainly by $\frac{V_2{}^2}{2}$ for given α of the guide vane canal.

The velocity triangle for Francis turbine shown in Fig. (**3.4**).

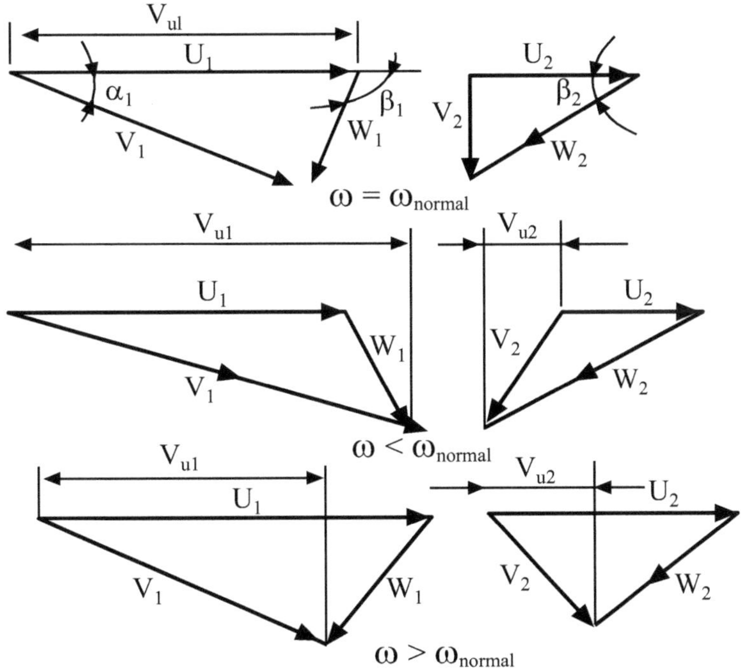

Fig. (3.4). Velocity triangle for three angular velocities [3].

w = w $_{normal}$ mean the rotational speed for which the turbine gives lowest energy loss at outlet represented mainly by $\frac{V_2{}^2}{2}$ and highest hydraulic efficiency for given angle α of the guide vane canal.

The velocity triangles for Kaplan turbine shown (Fig. **3.5**).

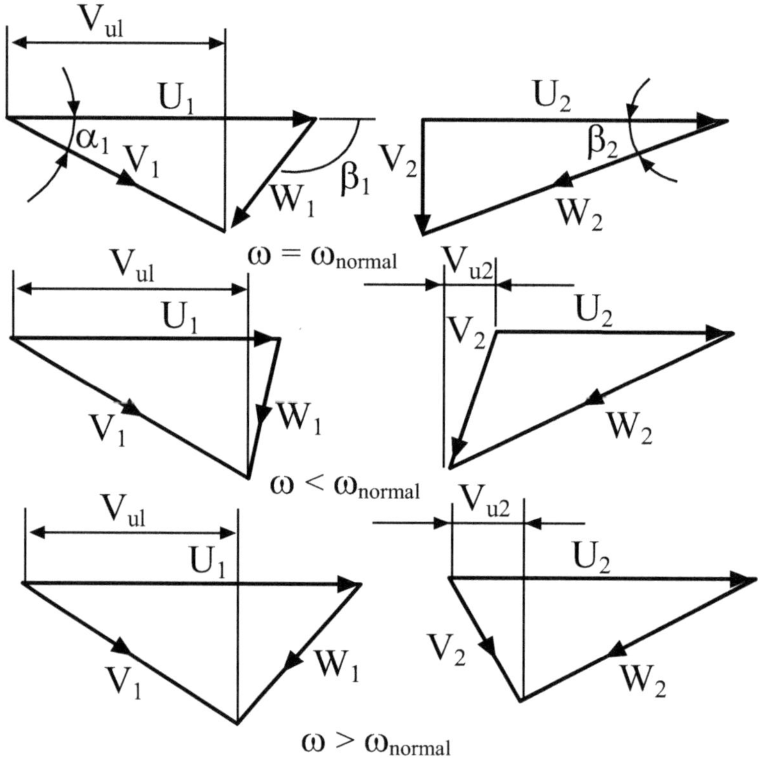

Fig. (3.5). Velocity triangle for three angular velocities [3].

For w = w $_{normal}$ the same as Frances turbine.

3.7. DRAFT TUBE

- To operate properly, reaction turbines must have a submerged discharge.
- The aim of the draft tube is also to convert the main part of the kinetic energy at the runner outlet to pressure energy at the draft tube outlet.
- This is achieved by increasing the cross section area of the draft tube in the flow direction.
- In an intermediate part of the bend, however, the draft tube cross sections are decreased in the flow direction to prevent separation and loss the efficiency.

If there is no draft tube, the kinetic head $\frac{{V_2}^2}{2g}$ would have been entirely lost.

$i.e.\ \Delta H = \frac{V_2^{\ 2}}{2g}$

Therefore, the kinetic head thus saved by draft tube, using the following sketch. (Fig. **3.6**).

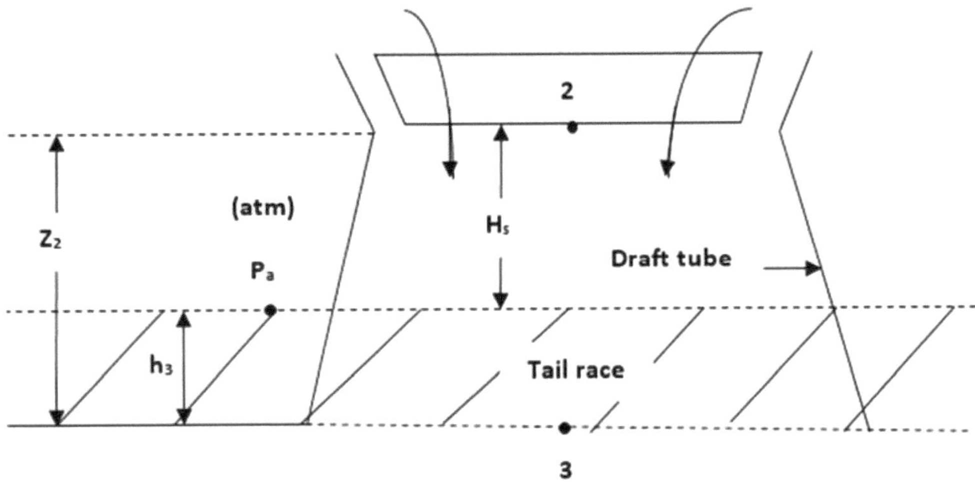

Fig (3.6). Sketch for draft tube cross – section.

Appling Bernoulli's equation between 2, 3 we get

$$\frac{P_2}{\gamma} + \frac{V_2^{\ 2}}{2g} + Z_2 = \frac{P_3}{\gamma} + \frac{V_3^{\ 2}}{2g} + Z_3 + h_f \qquad \textbf{(3.17)}$$

$$\therefore \frac{P_2}{\gamma} = \frac{P_3}{\gamma} + (Z_3 - Z_2) + \frac{V_3^{\ 2} - V_2^{\ 2}}{2g} + h_f$$

$$But\ \frac{P_3}{\gamma} = \frac{P_a}{\gamma} + h_3$$

$$where\ P_a = atmosphric\ pressure$$

$$And\ Z_2 - Z_3 - h_3 = H_s$$

$$(height\ of\ runner\ outlet\ above\ the\ tail\ race\ level).$$

$$\therefore \frac{P_2}{\gamma} = \frac{P_a}{\gamma} - \left(H_s + \frac{V_2{}^2 - V_3{}^2}{2g}\right) + h_f \quad \textbf{(3.18)}$$

$$And \ \frac{V_2{}^2 - V_3{}^2}{2g} = dynamic \ suction \ head$$

Minimum value of $\frac{P_2}{\gamma}$ is the vapor pressure at temperature of water T_2.

If we considering friction in draft tube as h_f which is in general expressed as a friction of dynamic suction head.

$$i.e. \ \ h_f = k\frac{V_2{}^2 - V_3{}^2}{2g}$$

Where k is the coefficient due to sudden expansion of the draft tube $= 1 - \left(\frac{A_2}{A_3}\right)^2$

$$\therefore \frac{P_2}{\gamma} = \frac{P_a}{\gamma} - \left(H_s + (1-k)\frac{V_2{}^2 - V_3{}^2}{2g}\right) \quad \textbf{(3.19)}$$

$$if \ \frac{P_2}{\gamma} < \frac{P_a}{\gamma} \ldots\ldots\ldots\ldots\ldots\ldots Cavitation \text{ occur}$$

From eq. (3.19) let $1 - k = \eta_d$ which is termed as the draft tube efficiency then

$$\frac{P_2}{\gamma} = \frac{P_a}{\gamma} - \left(H_s + \eta_d \frac{V_2{}^2 - V_3{}^2}{2g}\right) \quad \textbf{(3.19a)}$$

$$\text{and} \quad \eta_d = \frac{H_a - H_2 - H_s}{\frac{V_2{}^2 - V_3{}^2}{2g}} \quad \textbf{(3.19b)}$$

$$\text{or} \quad \eta_d = \frac{actual \ head \ converted}{theoretica \ \ head \ converted} \quad \textbf{(3.19c)}$$

3.8. NET HEAD

It is the total difference in elevation minus all losses (Fig **3.7**)

$$H_n = H_B - H_c$$

$$H_n = \frac{P_B}{\gamma} + \frac{V_B{}^2}{2g} + Z_B - \frac{V_c{}^2}{2g} \qquad Net\ head \tag{3.20}$$

$P_B = the\ pressure\ at\ inlet\ of\ the\ turbine$

$V_B = the\ velocity\ of\ flow\ at\ inlet\ of\ the\ turbine$

$V_c = the\ velocity\ at\ draft\ tube$ exit

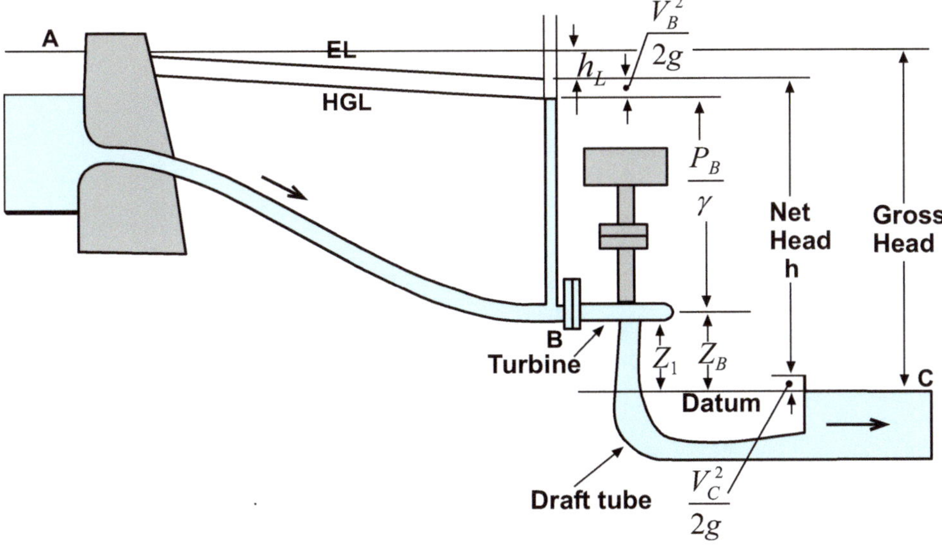

Fig. (3.7). Schematic diagram of hydraulic power plant [3].

Ex: A Franeis turbine is fitted with veritical shaped draft tube. The top and bottom diameters are equal to 60 cm and 90 cm respectively. The tube is running full with water flowing downwords and it has a vertical height of 6 m out of which 1.5 m is drowned in the tail race water. Assume friction losses of head between the top and the bottom points as 0.3 times the kinetic head at draft tube exit. The velocity at exit is 1.2 m/s.

Determine

a) the pressure head at the top point of the draft tube in m of water.

b) the total head at the same point with reference to the tail race as a datum.

c) the total head at the bottom point with reference to the tail race as a datum.

d) the power in the water at the top of the tube.

e) the power in the water at the bottom of the tube.

f) the efficiency of the draft tube.

Solution:-

a) Pressure head: $D_2 = 60\ cm\ ;\ D_3 = 90\ cm\ ; H = 6\ m$

$$H_L = 0.3\frac{{V_3}^2}{2g} \quad ; V_3 = 1.5\frac{m}{s}$$

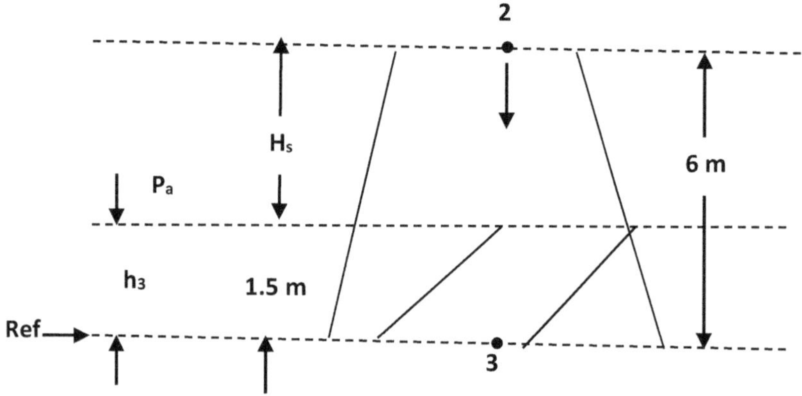

E.E. between 2,3

$$\frac{P_2}{\gamma} + \frac{{V_2}^2}{2g} + Z_2 = \frac{P_3}{\gamma} + \frac{{V_3}^2}{2g} + Z_3 + H_L$$

$$\frac{P_2}{\gamma} = \frac{P_3}{\gamma} - (Z_2 - Z_3) - \left(\frac{{V_2}^2 - {V_3}^2}{2g}\right) + 0.3\frac{{V_3}^2}{2g}$$

$$\frac{P_3}{\gamma} = \frac{P_a}{\gamma} + h_3$$

$$\frac{P_2}{\gamma} = \frac{P_a}{\gamma} - H_s - \left(\frac{{V_2}^2 - {V_3}^2}{2g}\right) + 0.3\frac{{V_3}^2}{2g}$$

From continuitg equation

$$A_2V_2 = A_3V_3 \quad \frac{\pi}{4}(0.6)^2 \times V_2 = \frac{\pi}{4}(0.6)^2 \times 1.5$$

$$V_2 = 3.38\, m/s$$

$$\therefore \frac{P_2}{\gamma} = 10 - 4.5 - \left(\frac{(3.38)^2 - (1.5)^2}{2 \times 9.81}\right) + 0.3\frac{(1.5)^2}{2 \times 9.81}$$

$$\frac{P_2}{\gamma} = 5.0654\, m\ of\ water\ absolute$$

$$or\ \frac{P_2}{\gamma} = -4.9346\, m\ of\ water$$

b) total head at (2)

$$H_2 = \frac{P_2}{\gamma} + \frac{{V_2}^2}{2g} + H_s = -4.9346 + \frac{(3.38)^2}{2 \times 9.81} + 4.5$$

$$H_2 = 0.1464\, m$$

c) at the bottom (3)

$$H_3 = \frac{P_3}{\gamma} + \frac{{V_3}^2}{2g} + Z_3 = H_2 - H_L$$

$$= 0.1464 - 0.3\frac{(1.5)^2}{2 \times 9.81} = 0.1112\, m$$

d) $P_2 = \gamma Q H_2 = 9.81 \times \left(\frac{\pi}{4}(0.6)^2 \times 3.38\right) \times 0.1464$

$$= 1.27\ kW$$

e) $P_3 = \gamma Q H_3 = 1.047\ kW$

f) $\eta_{draft} = \frac{P_{out}}{P_{in}} = \frac{P_3}{P_2} = \frac{1.047}{1.27}$

$$\eta_d = 82\%$$

3.9. WORKING PROPERTIES OF REACTION TURBINES

The exact shapes of reaction turbine runner depends on its specific speed. It is obvious from the equation of specific speed (refer to eq. (3.6)) that higher specific speed means lower head. This requires that the runner should admit a comparatively large quantity of water. This can done by increasing either:

(i) The Breadth ratio $= \frac{B}{D}$

Where B – the guide vane width or depth and

D – the diameter of the runner.

(ii) The speed ratio $\phi = \frac{U}{\sqrt{2g}}$

Where U the tangential velocity of the runner.

(iii) the flow ratio $\varphi = \frac{V_f}{\sqrt{2gH}}$

Where V_f – the radial velocity of flow

In the first case the width of guide vane B and correspondingly the height of the runner should increase. In the second case the head should increase while in the third case V_f the radial velocity of flow through the guide vanes should increase.

The practical range of the speed, flow and breadth ratios for Francis and Kaplan runner design could be found as follows.

1. Francies Turbine:

$- \text{Speed ratio} = \frac{U}{\sqrt{2g}} \quad its\ value\ \ Renerally\ between\ 0.6\ \rightarrow\ 0.9$

$- \text{Flow ratio} = \frac{V_f}{\sqrt{2gH}} \quad its\ value\ \ 0.15 \rightarrow 0.3$

$- \text{Breadth ratio} = \frac{B}{D} \quad its\ value\ \ 0.15 \rightarrow 0.4$

2. Kaplan Turbines:

$\text{Speed ratio} \quad upto \rightarrow 2.09$

$\text{Flow ratio} \quad upto \rightarrow 0.69$

$\text{Breadth ratio} \ \ upto \rightarrow 0.35 \rightarrow 0.6$

EX. 1: A reaction turbine has outer and inner diameters of the wheel as (1 and 0.5) meters respectively. The vanes are radial at inlet and the discharge is radial at outlet and the enters the vanes at an angle of 10°. Assuming the velocity of flow as constant and equal to 3 m/s. find the speed of the wheel and the vane angle at outlet.

Solution:-

$D_1 = 1\,m\,;\ D_2 = 0.5\,;\ \beta_1 = 90°\ (vane\ radial)$

$\alpha_2 = 90°(discharge\ radial\,;\ \alpha_1 = 10°\,;\ V_{f1} = V_{f2} = 3\ m/s$

The velocity diagram at inlet and outlet

$$U_1 = \frac{V_{f1}}{\tan\alpha_1} = \frac{3}{\tan 10°} = 17\ m/s$$

$$also\ U_1 = \frac{\pi D_1 N}{60}$$

$$\therefore N = \frac{17\times60}{\pi\times1}$$

$$N = 325\ rpm$$

At outlet

$$U_2 = \frac{\pi D_2 N}{60} = \frac{\pi\times0.5\times325}{60} = 8.5\ m/s$$

Then $\tan\beta_2 = \frac{V_{f2}}{U_2} = \frac{3}{8.5} = 0.353$

$or\ \beta_2 = 19.27°$

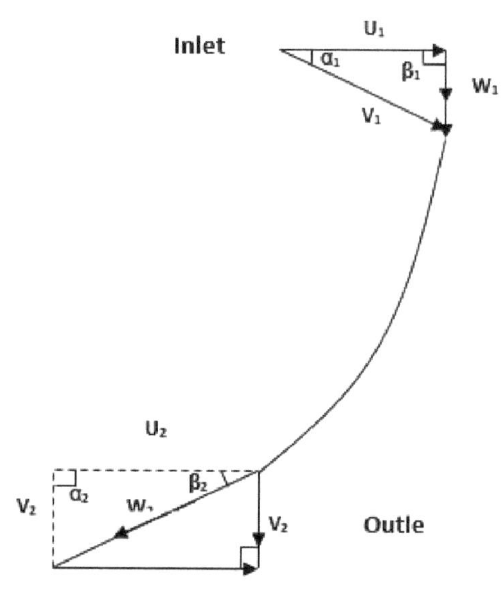

EX. 2: A reaction turbine with a supply of 550 l/s under a head of 15 m develops 73.6 kW at 375 rpm. The inner and the outer diameter of the runner are 50 cm and 75 cm respectively. The velocity of water at exit is 3 m/s. Assuming that the discharge is radial and that the width of the wheel is constant. Find the actual and the theoretical hydraulic efficiency of the turbine and the inlet angles of the guide and wheel vanes.

Solution:-

$Q = 0.55\ m^3/s\ ; D_1 = 75\ cm\ ;\ D_2 = 50\ cm$

$H = 15\ m\ ; P_t = 73.6\ kW\ ;\ V_{f2} = 3\ m/s\ ; N = 375\ rpm$

$\alpha_2 = 90°\ (radial\ discharge)\ B_1 = B_2\ width\ of\ runner.$

a) $U_1 = \frac{\pi D_1 N}{60} = \frac{\pi\times0.75\times375}{60} = 14.75\ m/s$

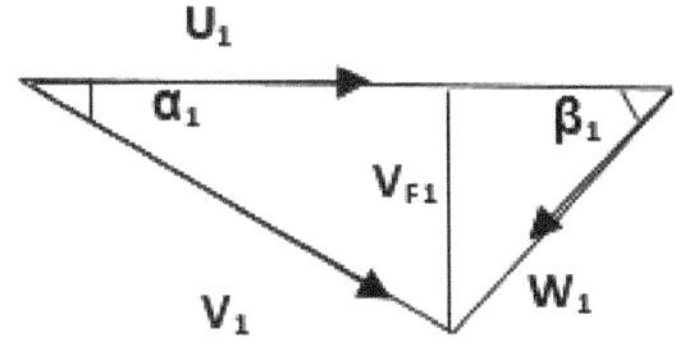

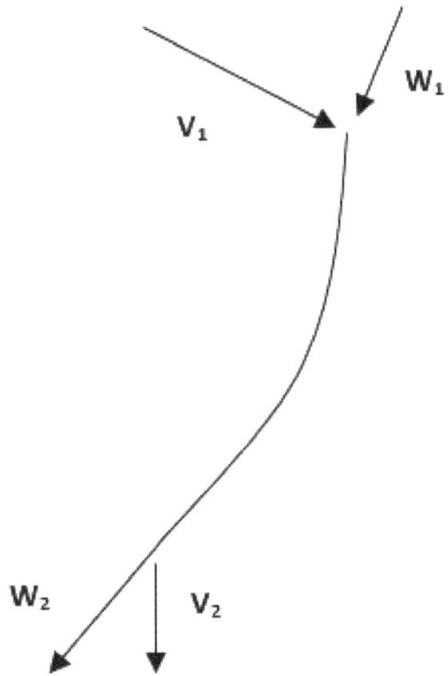

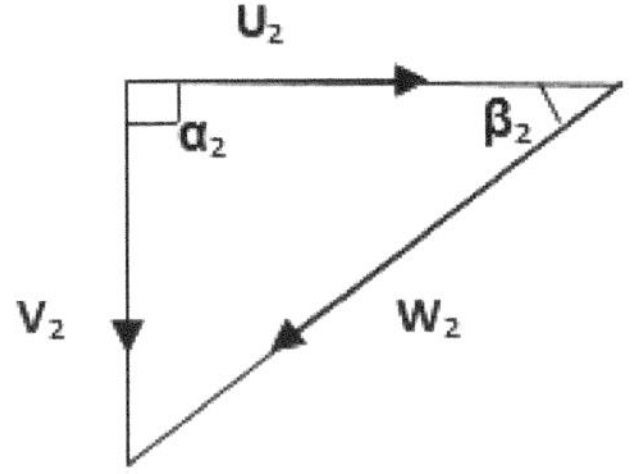

From velocity triangle at exit

$$V_{f2} = 3\ m/s = V_2\ ;\ \alpha_2 = 90°$$

Work done /sec by the turbine per kg of water $= \frac{V_{u1}U_1}{g}$

And this is equal to the head utilized by the turbine.

$\frac{V_{u1}U_1}{g} = H - \frac{V_2{}^2}{2g}$ (Assuming no pressure head loss at outlet)

$$Or\ \frac{V_{u1} \times 14.75}{9.81} = 15 - \frac{3^2}{19.62}$$

$$\therefore\ V_{u1} = 9.7\ m/s$$

$\therefore$ The power developed by the turbine runner $= \rho Q V_{u1} U_1$

$$= 1000 \times 0.55 \times 9.7 \times 14.75$$

$$= 78.7\ kW$$

$$\text{The available power of water } = \gamma QH = 9.81 \times 0.55 \times 15$$

$$= 80.93\ kW$$

$$\therefore\ overall\ efficiency\ \eta_o = \frac{P_{out}}{P_{in}} = \frac{73.6}{80.93} = 90.9\%$$

$$and\ hydraulic\ ficiency\ \eta_h = \frac{78.7}{80.73} = 97.24\%$$

b) $Q = (\pi D_1 B_1)V_{F1} = \pi D_2 B_2 V_{F2}$

$$\therefore\ V_{f1} = V_{f2} \times \frac{D_2}{D_1} = 3 \times \frac{0.5}{0.75} = 2\ m/s$$

From inlet velocity triangle

$$\tan\beta_1 = \frac{V_{F1}}{U_2 - V_{u1}} = \frac{2}{14.75 - 9.7}$$

$$\beta_1 = 21.6°$$

$$Then\ V_1 = \sqrt{{V_{u1}}^2 + {V_{F1}}^2} = \sqrt{(9.7)^2 + (2)^2} = 9.9\ m/s$$

$$and\ V_{u1} = V_1 \cos\alpha_1 \quad \therefore\ \cos\alpha_1 = \frac{V_{u1}}{V_1} = \frac{9.7}{9.9}$$

$$\alpha_1 = 11.54°$$

EX.3: A reaction turbine rotates at 370 rpm. The wheel vanes are radial at inlet and $D_1 = 2D_2$. The constant velocity of flow in the wheel 2 m/s, water enters the wheel at an angle of (10.07°) to the tangent of the wheel at inlet. The breadth of the wheel at inlet is 75 mm and the area of flow blocked by the vanes is 5% of the gross area of flow at inlet. Find

a) The outer and inner diameters of the wheel.

b) The net available head.

c) The wheel vane angle at outlet.

d) The theoretical power developed by the wheel.

Solution:-

$N = 370\, rpm\, ;\; D_1 = 2D_2\, ;\; V_{f1} = V_{f2} = 2\, m/s$

$\alpha_1 = 10.07°\, ;\; B_1 = 75\, mm$ Contraction coefficient K=1-0.5=0.95

$\beta_1 = 90°$ (Vanes are radial at inlet)

a) $V_{f1} = V_1 \sin\alpha_1 \quad \therefore\; V_1 = \frac{V_{f1}}{\sin\alpha_1}$

$$V_1 = 11.43\, m/s$$

$$V_{u1} = V_1 \operatorname{coc}\alpha_1 = 11.25\, m/s$$

$$Assume\; \beta_1 = 90° \quad \therefore\; V_{u1} = U_1$$

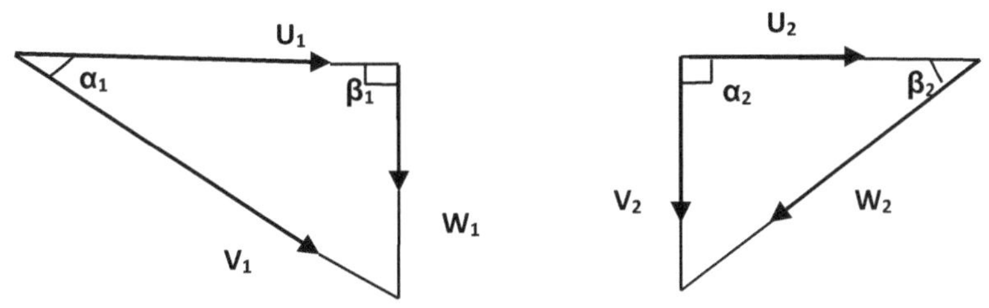

$$\text{And } U_1 = \frac{\pi D_1 N}{60} = \frac{\pi D_1 \times 370}{60} = 11.25$$

$$\therefore\; D_1 = 0.58\, m\;\; and\; D_2 = 0.29\, m$$

b) Net available head at the wheel =Work done /sec per unit mass of water

$$H = \frac{V_{u1}U_1 - V_{u2}U_2}{g}$$

$$or\ H = \frac{V_{u1}U_1}{g}\ for\ radial\ outlet\ flow\ \alpha_2 = 90°$$

$$H = \frac{11.25 \times 11.25}{9.81} = 12.9$$

c) $$U_2 = \frac{\pi \times 0.29 \times 370}{60} = 5.63\ m/s$$

$$V_{F2} = 2\ m/s$$

$$\therefore\ \tan\beta_2 = \frac{V_{F2}}{U_2} = \frac{2}{5.63}\ m/s$$

$$\beta_2 = 19.5°$$

d) $Q = \pi D_1 B_1 V_F$ Contraction coefficient (K)

$$= \pi \times 0.58 \times 0.075 \times 2 \times 0.95$$

$$= 0.26\ m^3/s$$

$$\therefore\ available\ power\ P_a = \gamma QH$$

$$= 9.81 \times 0.26 \times 12.9$$

$$= 32.9\ kW$$

EX. 4: A Francis turbine is required to developed 3680 kW when operating under a net of 30 m and the specific speed to be about 231.55 m-kW unit assuming guide vane angle at full gate opening 30° hydraulic efficiency 90%, overall efficiency 87%, radial velocity of flow at inlet $0.3\sqrt{2gH}$, blade thickens coefficient 5% draw the inlet velocity diagram and find: -

a) The nearest synchronous speed to drive an alternator to give a frequency of 50 cycles per sec.

b) The diameter and the width of the runner at inlet.

c) The theoretical inlet angle of the runner vanes.

Solution:-

$$P_t = 3680\ kW\ ;\ \alpha_1 = 30°\ ; H = 30\ m$$

$$V_{F1} = 0.3\sqrt{2gH}\ ;\ N_s = 231.55\ ; K = 0.95$$

$$\eta_h = 0.9\ ; f = 50\ cycles/sec\ ;\ \eta_t = 0.87$$

a) $N_s = \frac{N\sqrt{P_t}}{H^{\frac{5}{4}}} = \frac{N\sqrt{3680}}{(30)^{\frac{5}{4}}} = 231.55 \therefore N = 268\ rpm$

Assuming No. of poles 11

$\therefore$ Nearest synchronous speed $= \frac{3000}{11} = 272.8\ rpm$

b) Velocity of flow $V_{F1} = 0.3\sqrt{2 \times 9.81 \times 30} = 7.26\ m/s$

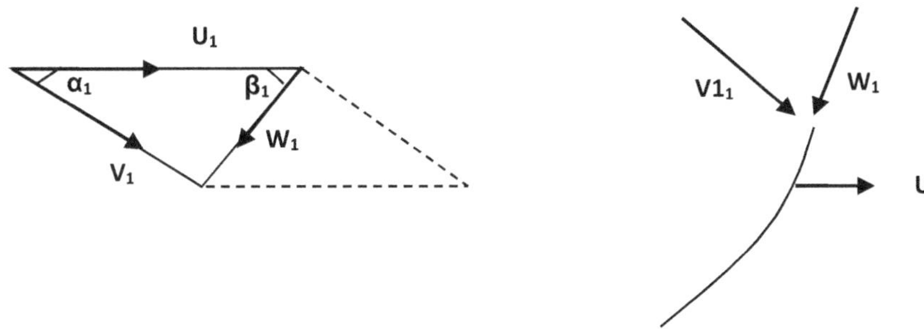

$$V_{F1} = V_1 \sin \alpha_1 \quad \therefore V_1 = \frac{7.26}{\sin 30} = 14.52\ m/s$$

$$and\ V_{u1} = V_1 \cos \alpha_1 = 14.52 \times \cos 30 = 12.6 \frac{m}{s}$$

$$\therefore \eta_h = \frac{V_{u1} U_1}{gH} \quad \therefore H = \frac{V_{u1} U_1}{g \eta_h}$$

$$U_1 = \frac{30 \times 0.9 \times 9.81}{12.6} = 21\ m/s$$

$$and\ U_1 = \frac{\pi D_1 N}{60} \quad D_1 = \frac{60 \times 21}{\pi \times 272.8} = 1.47\ m$$

$$Now\ P_t = \gamma Q H \quad \therefore Q = \frac{3680}{9.81 \times 30} = 14.4\ m^3/s$$

$$Q = \pi D_1 B_1 V_{F1}.K$$

$$14.4 = \pi \times 1.47 \times B_1 \times 7.26 \times 0.95$$

$$B = 0.453\ m\ the\ width\ of\ the\ runner$$

c) $\tan \beta_1 = \frac{V_{F1}}{U_1 - V_{u1}} = \frac{7.26}{21-12.6} = 0.865$

$$\therefore \beta_1 = 40.6°\ the\ inlet\ angle\ of\ the\ vane$$

EX. 5: A reaction turbine, the peripheral velocity of the wheel at inlet is given by $U_1 = \emptyset\sqrt{2gH}$ and the radial velocity of flow $V_F = \varphi\sqrt{2gH}$. The breadth of the wheel at inlet is n' time the diameter of the runner at inlet. If the turbine efficiency is 85% and the area taken up by vanes at inlet 5% of the peripheral area at inlet. Prove that the specific speed of the turbine is equal to $(888.2\emptyset\sqrt{\varphi n})$. A runner of the above type having a specific speed of 188.7 m-kW unit required to develop 6625 kW under a head 85 m. taken φ as 0.18 and n=0.2, calculate the rpm and the diameter of the runner.

Solution:-

$$\varphi = 0.18\ ;\ B_1 = nD_1 \ \therefore\ B_1 = 0.2D_1$$

$$\eta_t = 0.85\ ;\ P_t = 6625\ kW\ ;\ N_s = 188.7$$

$$H = 85\ m\ ; K = 0.95$$

a) $Q = 0.95 \times \pi \times D_1 \times B_1 \times V_{F1}$

$$= 0.95 \times \pi \times D_1 \times nD_1 \times \varphi\sqrt{2gH}$$

$$= 121.88 \times n \times {D_1}^2 \times \varphi \ldots\ldots\ldots (1)$$

$$U_1 = \frac{\pi D_1 N}{60} = \emptyset\sqrt{2gH}$$

$$N = \frac{60 \times \emptyset\sqrt{2 \times 9.81 \times 85}}{\pi D_1} = \frac{780\emptyset}{D_1} \dots\dots\dots (2)$$

$$Now\ N_s = \frac{N\sqrt{P_t}}{H^{\frac{5}{4}}} \quad P_t = \gamma QH\eta_t$$

$$= 9.81 \times 121.88 \times nD_1{}^2\varphi \times 0.85 \times 85$$

$$\therefore P_t = 86385.2\ nD_1{}^2\varphi \dots\dots\dots (3)$$

$$\therefore N_s = \frac{\frac{780\emptyset}{D_1} \times \sqrt{86385.2nD_1{}^2\varphi}}{(85)^{\frac{5}{4}}}$$

$$N_s = \frac{\frac{780\emptyset}{D_1} \times D_1 \times 294\sqrt{n\varphi}}{258.1} = (888.2\emptyset\sqrt{\varphi n})$$

$$\therefore N_s = (888.2\emptyset\sqrt{\varphi n})$$

b) $N_s = \frac{N\sqrt{P_t}}{H^{\frac{5}{4}}}$

$$188.7 = \frac{N \times \sqrt{6625}}{85^{1.25}}$$

$$N = 600\ rpm$$

$$\therefore Q = 0.95 \times \pi \times D_1 \times B \times V_F$$

$$Q = 0.95 \times \pi \times D_1{}^2 \times n \times \varphi\sqrt{2gH}$$

$$= \frac{P_t}{\gamma H \eta_t}$$

$$since\ n = 0.2\ \varphi = 0.18$$

$$\therefore\ 0.95 \times \pi \times {D_1}^2 \times 0.2 \times 0.18 \times \sqrt{2 \times 9.81 \times 85}$$

$$= \frac{6625}{2 \times 9.81 \times 85}$$

$$D_1 = 1.46\ m$$

EX. 6: A Kaplan turbine develops 5900 kW under an effective head of 5 m. Its speed ratio is 2 and flow ratio is 0.6 and the diameter of the boss=0.35 times the external diameter of the runner. Overall efficiency of the turbine is 90%. Calculate the diameter of the runner and also the specific speed.

Solution:-

$$P_t = 5900\ kW\ ; H = 5\ m\ ;\ \eta_t = 90\%$$

$$\emptyset = 2\ ; \varphi = 0.6\ ; d = 0.35 D_1$$

$$\eta_t = \frac{P_{out}}{P_{in}} = \frac{P_t}{\gamma Q H} = \frac{5900}{9.81 \times Q \times 5} = 0.9$$

$$Q = 133.3\ m^3/s$$

$$V_F = \varphi\sqrt{2gH} = 0.6\sqrt{2 \times 9.81 \times 5}$$

$$= 5.92\ m/s$$

$$Q = AV_F \quad \therefore A = \frac{Q}{V_F}$$

$$area\ of\ flow = \frac{133.3}{5.92} = 22.5\ m^2$$

$$= \frac{\pi}{4}{D_1}^2 - \frac{\pi}{4}{d_1}^2 = \frac{\pi}{4}({D_1}^2 - (0.35D_1)^2)$$

$$\therefore 22.5 = \frac{\pi}{4}{D_1}^2(1 - (0.35)^2)$$

$$\therefore\ D_1 = 5.7\ m$$

$$and\ U_1 = \frac{\pi D_1 N}{60} = 2\sqrt{2gH}$$

$$\therefore\ \frac{\pi \times 5.7 \times N}{60} = 2\sqrt{2 \times 9.81 \times 5}$$

$$N = 66.4\ rpm$$

$$\therefore N_s = \frac{N\sqrt{P_t}}{H^{\frac{5}{4}}}$$

$$= \frac{66.4\sqrt{5900}}{6^{\frac{5}{4}}}$$

$$N_s = 682\ m - kW\ units$$

3.10. POWER REGULATING MECHANISMS

1- Guide – Vane Operating Gear: -

The operating gear must ensure setting of all guide vanes at the same angle (equal values of α_1) with any openings of the wicket gate.

The schematic of the most widely employed guide: - Vane operating gears is show in Fig (**3.8**). The levers 1 set on the upper pivot of the guide vanes are connected to the regulating ring 3 by means of shackles and pull rods 2. These three elements constitute the main part of the guide - vane operating gear (Fig. **3.9a**). Shows the guide-vane operating gear in the full closing position. If the regulating ring turns counterclockwise, all levers will turn through the same angle, as well as the guide vanes and the turbine will be opened (Fig. **3.9b**). It follows that to change the turbine power it is necessary to turn the regulating ring. Servomotors of the wicket gate are designed for turning the regulating ring and changing the opening of the guide vanes so as to regulate turbine power. The servomotors are operated by high pressure oil and they find use in all regulating system of high-power turbine.

The servomotors may be operating the guide vanes of the turbine as presented schematically in Fig. (**3.10**) with two cylinders. There are many designed for servomotors application as shown in Figs. (**3.11**, **3.12**).

2- Runner- Blade Operating Gear of Adjustable-Blade: -

The runner operating gear must ensure a change in the blade angle and accurate setting of the blades as required without stopping the turbine, *i.e.* with a rotating runner. The blade operating rings must be able to overcome tremendous forces caused by water pressure, centrifugal and frictional forces in the blade pin bearings, they must arrange in very confined space in the runner hub and feature an exceptionally high reliability, since their repair requires complete disassembly of units.

The schematic diagram of a runner blade operating gear is shown in Fig. (**3.12**). Blade pivot 2, having two supports in housing 3 and 4 is connected to the blade flange 1. The lever 5 set on pivot 2 is connected by pull rod 6 to piston 7 of the runner servomotor. Displacement of the piston 7 through pull rod 6 results in turning of the pivot 2 and blade 1.

There exist a great number of designs of the above considered runner blade operating, mainly differing in the way in which the servomotor piston is connected to the levers. Two examples are shown in Figs (**3.13** and **3.14**) for high-head and law-head axial flow turbines [6].

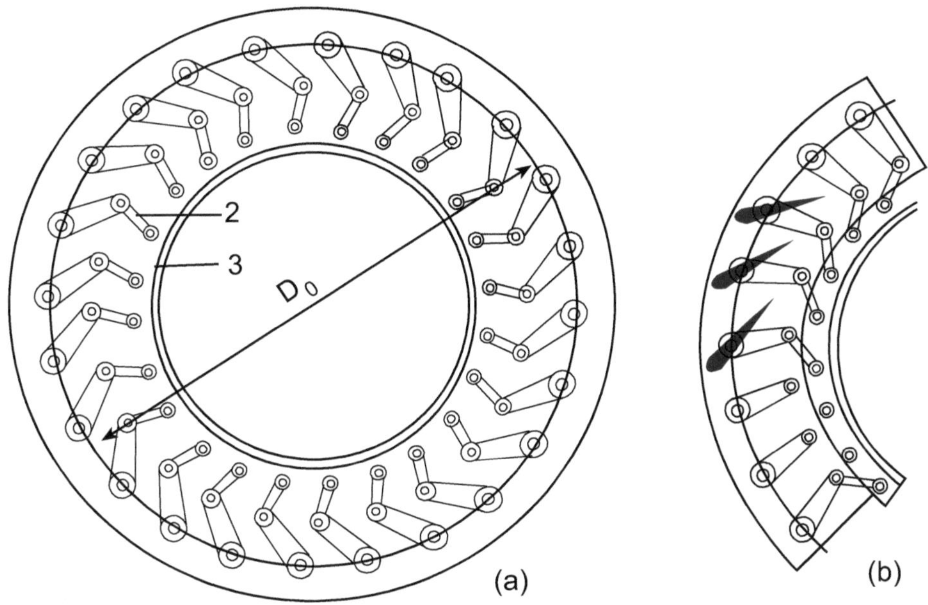

Fig. (3.8). Guide vane operating mechanism. Schematic diagram [6].

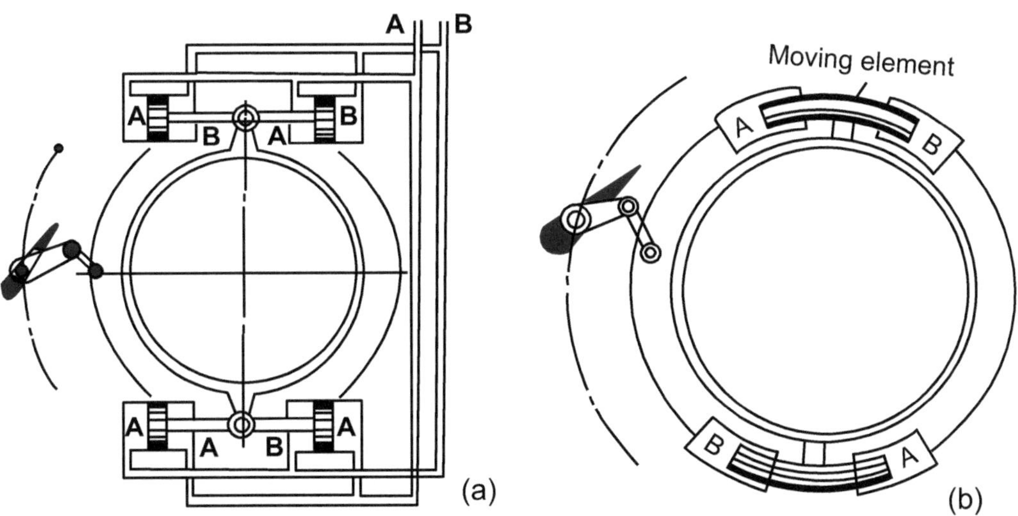

Fig. (3.9). Schematic diagram of guide vane operating gear with two twinned (**a**) and anchor-ring (**b**) servomotors [6].

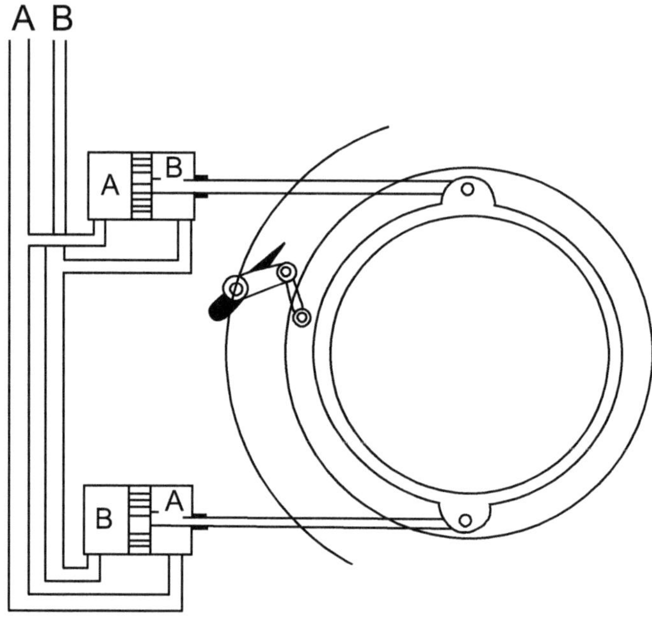

Fig. (3.10). Schematic diagram of guide vane operating gear with two cylindrical servomotors [6].

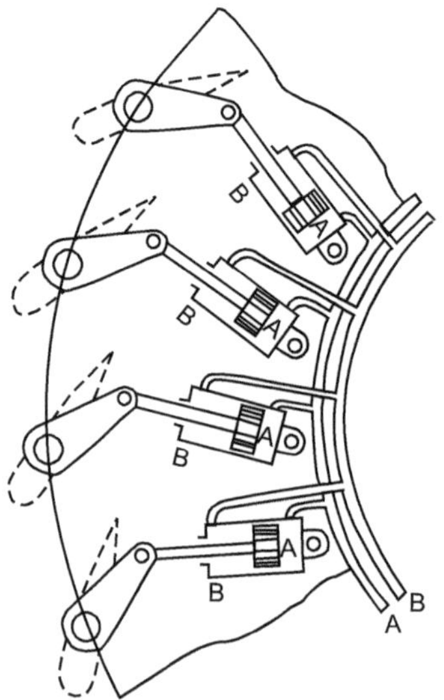

Fig. (3.11). Schematic diagram of guide vane operating gear with individual servomotors [6].

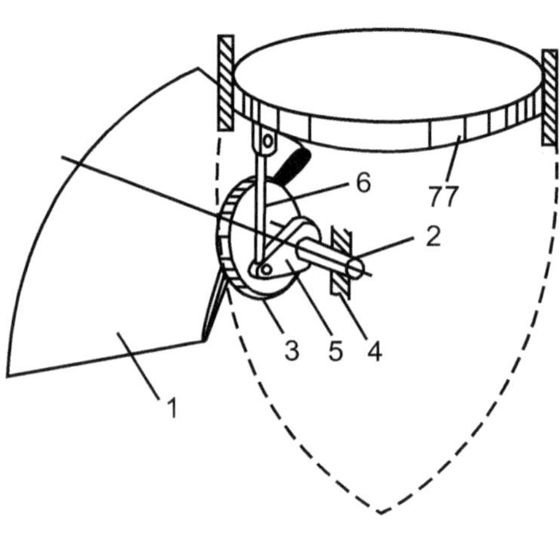

Fig. (3.12). Schematic diagram of adjustable-blade runner operating gear [6].

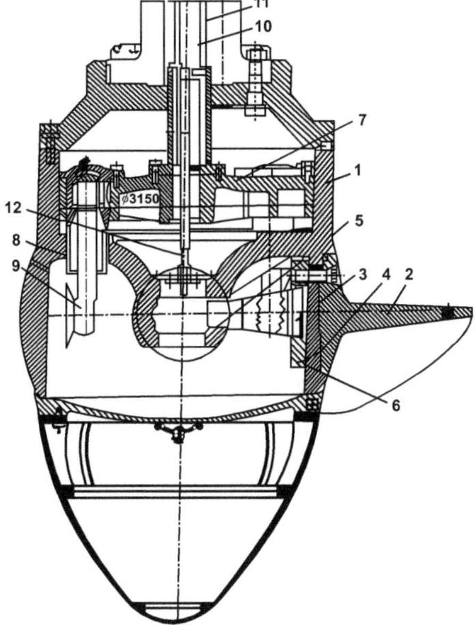

Fig. (3.13). Blade operating gear of high-head turbine run [6].
1. Rinner bearing 2. Blade 3. Point with flange 4,5. Plain bearing 6. Lever 7. Servommeter piston 8. Cups rigidly connected to piston 7 9. Pull rod and hinging the cup and piston with pin of the blade lever. 10,11. Oil supply pipes to servometer. 12. Middle pipe serves to drain the oil.

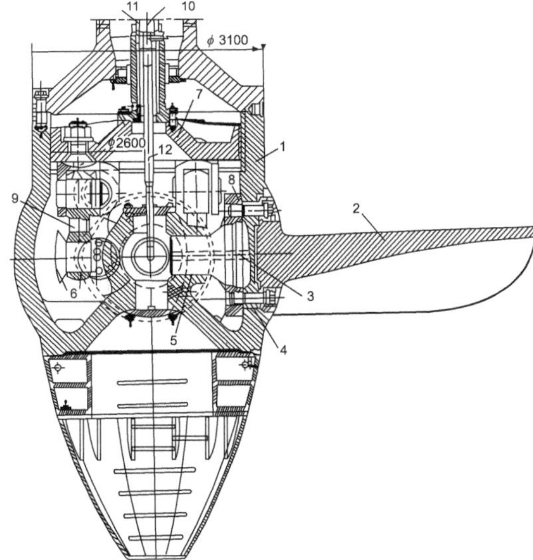

Fig. (3.14). Blade operating gear of low-head turbine runner [6].
1. hub 2. Blades 3. Pivot flange 4,5. Plain bearings 6. Blade levers 7. Piston 8. Cups rigidly connected to piston 9. Pull rod 10, 11. Oil supply pipes 12. Middle pipe serves to drain oil.

3.11. SUPPLY AND DISCHARGE SYSTEMS

1- Penstock:- in determining the number of penstocks for any particular installation various factors have to be considered. Let us compare by a single penstock and by a system of n realized by selecting diameters either, (Fig. **3.15**)

a) for identical flow velocities.

b) for identical friction losses.

Let Q: Discharge conveyed in a single penstock.

D: Diameter of the penstock.

V: flow velocity.

h_c: Head loss.

e: wall thickness.

G: weight of the penstock.

a) Identical Flow Velocities

Dividing the discharge Q among n conduits, the diameter of each pipe should be determined to ensure an identical flow velocity V. with each penstock discharging.

$$Q_n = \frac{Q}{n}$$

$$for\ identical\ velocity\ V = \frac{Q}{\frac{\pi D^2}{4}} = \frac{Q_n}{\frac{\pi D_n{}^2}{4}} \quad \textbf{(3.21)}$$

Where D_n is the diameter of any penstock:

From eq. (3.21)

$$\therefore D_n = D\sqrt{\frac{Q_n}{Q}} = \frac{D}{\sqrt{n}} \quad \textbf{(3.22)}$$

The head loss due to friction in case of single penstock installation:

$$h_f = f.\frac{L}{D}.\frac{V^2}{2g} \qquad since\ Q = AV$$

$$or\ h_f = \frac{8f}{g\pi}.\frac{Q^2}{D^5} \tag{3.23}$$

$$let\ \frac{8fL}{g\pi} = a_1 = 0.26\ fL$$

$$\therefore h_f = a_1\frac{Q^2}{D^5} \tag{3.24}$$

And for n penstocks

$$h_{fn} = a_1\frac{\left(\frac{Q}{n}\right)^2}{\left(\frac{D}{\sqrt{n}}\right)^5} = a_1\frac{Q^2\sqrt{n}}{D^5} \tag{3.25}$$

$$\therefore h_{fn} = h_f\sqrt{n} \tag{3.26}$$

The wall thickness in case of single penstock arrangement:

$$e = \frac{PD}{2\sigma_{steel}} \rightarrow a_2 = \frac{P}{2\sigma_{steel}}$$

$$e = a_2D \tag{3.27}$$

$$P = static\ + water\ hammer\ pressure$$

$$\sigma = tensile\ stress\ of\ the\ steel$$

$$For\ n\ penstocks,$$

$$e_n = a_2D_n = a_2\frac{D}{\sqrt{n}}$$

$$\therefore e_n = \frac{e}{\sqrt{n}} \quad \textbf{(3.28)}$$

The total penstock weight in case of single penstock installation

$$G = \gamma_{steel} \pi D e \quad a_3 = \gamma_{steel} \pi$$

$$\therefore G = a_3 D e \quad \textbf{(3.29)}$$

Also for n penstocks

$$G_n = \frac{G}{n} \quad \textbf{(3.30)}$$

Or the total weight of n penstocks

$$n G_n = G$$

b) Identical Friction Head Losses

for determining the diameter D_n ensuring a head loss identical with that in the single penstock,

$$h_f = a_1 \frac{Q^2}{D^5} = a_1 \frac{\left(\frac{Q}{n}\right)^2}{{D_n}^5}$$

$$D_n = \frac{D}{\sqrt[5]{n^2}} \quad \textbf{(3.31)}$$

Also wall thickness of the n penstocks

$$e_n = \frac{e}{\sqrt[5]{n^2}} \quad \textbf{(3.32)}$$

And the weight $G_n = \frac{G}{\sqrt[5]{n^4}}$ **(3.33)**

As can be seen the theoretical weight increase for several penstocks with $n^{1/5}$ on the other hand it is more safety of operation when use n penstocks.

Large power penstocks subject to heads of several hundred meters may be constructed of banded steel pipe.

Simple steel pipes are used for

$$PD < 10000\ (kg/cm)$$

Banded steel pipe for

$$PD < 10000\ (kg/cm)$$

$$Where\ P: kg/cm^2\ D: pipe\ diameter\ (cm)$$

Practical empirical equations used to find out the diameter of a penstock will give (Fig. **3.15**).

For maximum velocity in the penstock may be V_{max}=6 m/s

$$h_f = \frac{V^2 L' n^2}{R^{\frac{4}{3}}} \leq 0.05\ H_{gross} \quad \textbf{(3.34)}$$

$$and\ H_{gross} < 100\ m\ D = \sqrt[7]{0.05Q^3}\ m \quad \textbf{(3.35)}$$

$$H_{gross} > 100\ m\ D = \sqrt[7]{\frac{5.2Q^3}{H_{gross}}}\ m \quad \textbf{(3.36)}$$

$$R = hydraulic\ radius$$

$$n = manning\ coefficient$$

$$L' = penstock\ length\ (m)$$

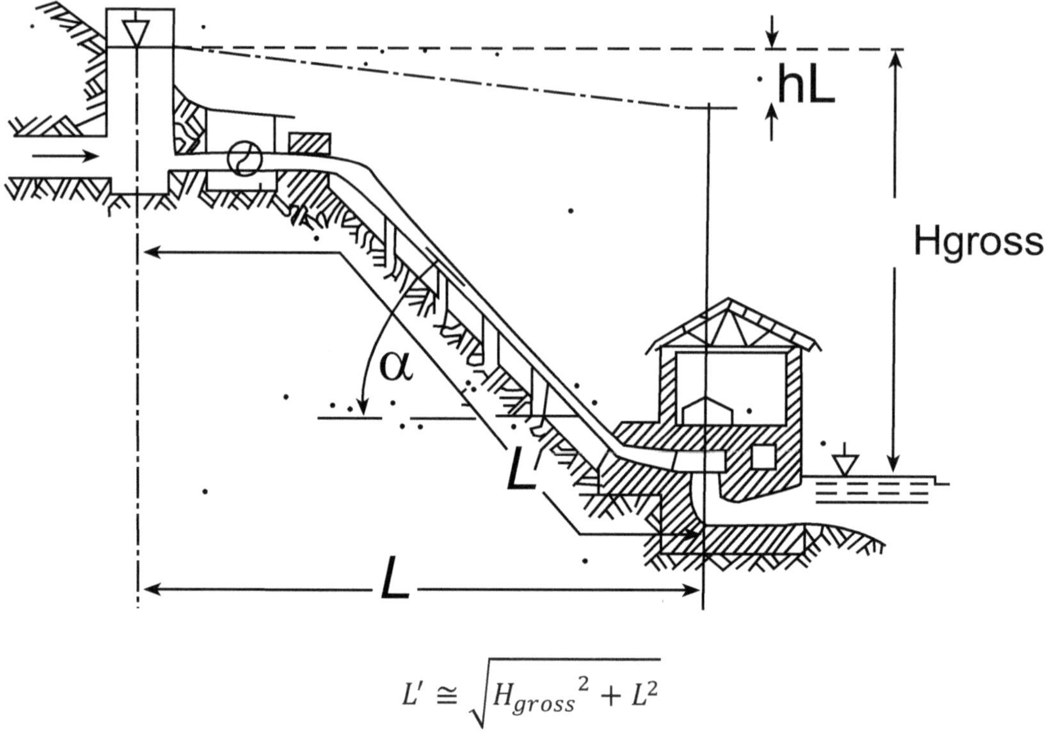

$$L' \cong \sqrt{H_{gross}{}^2 + L^2}$$

Fig. (3.15). Schematics diagram of power plant [3].

2- Turbine Case

The turbine case serves to supply water to the guide vanes (wicket gate) of a reaction turbine.

Turbine cases must satisfy the following requirements:

1- They must ensure uniform supply of water to the guide vanes over the entire perimeter of the wicket gate.

2- They must ensure minimum hydraulic losses in the case proper, in the stayring and where water flow enters the guide vanes.

3- The shape and size of a turbine case must be in agreement with the layout of the power house of the HEPP.

3- Draft Tubes

In reaction turbine water is discharged from the runner into a draft tube through which the water is discharged into tail water pool. The draft tube affects materially the power-generating properties of turbines, especially of low-head ones. In addition, the draft tube determines the dimensions of the lower part of the power house of HEPP and the elevation of the base. In this convection, great attention is devoted to the definition of the form and dimension of draft tube in designing HEPP.

Two types of draft tube are distinguished straight and elbow type:

a- Straight Draft Tube:

The simplest is the straight-type conical draft tube (Fig. **3.16**) that possesses good power-generating characteristics; however, it must be of a considerable required length issue. For high-power vertical turbine this makes it necessary to construct the base at a considerable depth and results in a greater cost at the HEPP. In this connection, draft tubes of this type find application only in small-power turbines.

b- Elbow-type Draft Tubes:

It is used in almost all HEPP where high-power vertical turbines operate.

In order to keep down the cost of excavation, particularly in rock, the vertical length of the draft tube should be minimum, (Fig. **3.17**). Since the draft tube exit diameter should be as large as possible to recover the kinetic head and the same time the maximum value of the cone angle is fixed, the draft tube must be bent to keep its, definite length.

Fig. (**3.18**) illustrate an Elblow Tube with circular inlet and a rectangular outlet section. This type of draft tube has been designed to turn the water from the vertical to horizontal direction with minimum depth of excavation and at the same time having high efficiency.

A comparison between impulse and reaction turbine shown in Table **3.1**.

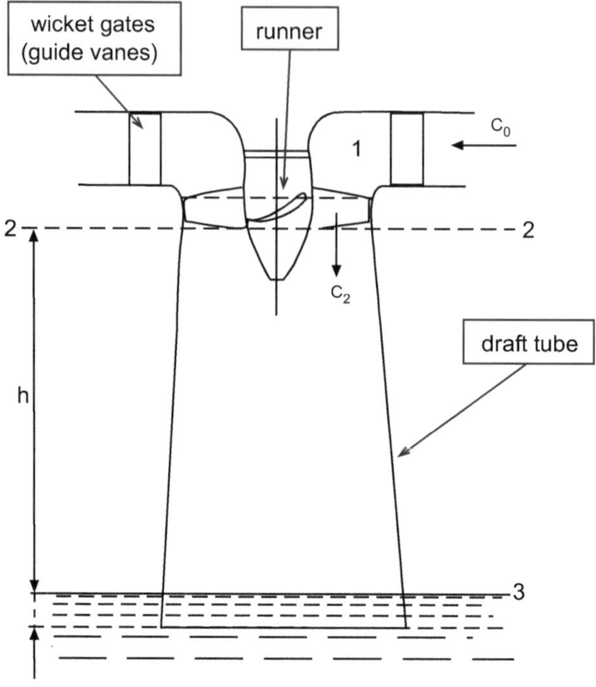

Fig. (3.16). Straight draft tube [4]. V_0, V_2 – The velocity of flow at inlet and outlet of the runner respectively.

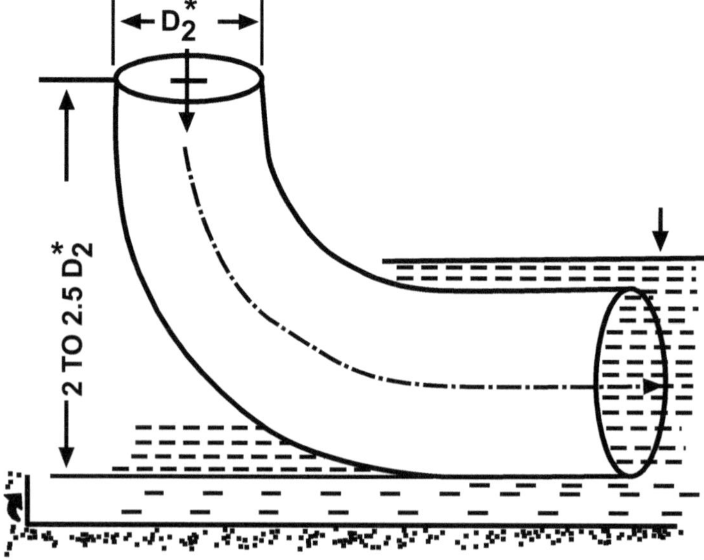

Fig. (3.17). Schematic diagram of axial-flow turbine with draft tube [4].

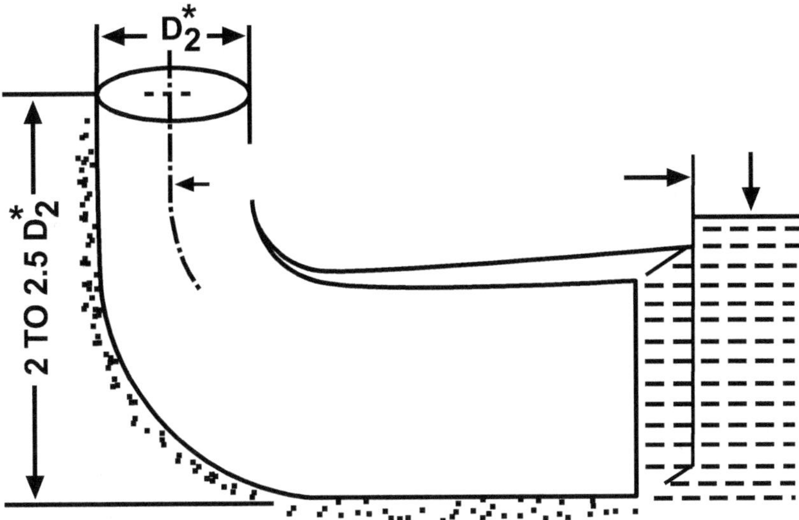

Fig. (3.18). Standard elbow - type draft tube [4].

Table 3.1. A comparison between impulse and reaction turbine.

Impulse Turbine	**Reaction Turbine**
The entire available energy of the water is converted into kinetic energy.	Only a portion of the energy is converted into kinetic energy before the fluid enters the turbine runner.
The work is done only by the change in the Kinetic energy of the jet.	The work is done partly by the change in the velocity head, but almost entirely by the change in pressure head.
Flow regulation is possible without loss.	It is possible to regulate the flow without loss.
Unit is installed above the tailrace.	Unit is entirely submerged in water below the tailrace.
Casing has no hydraulic function to perform, because the jet is unconfined and is it atmospheric. Thus, casing serves only to prevent splashing of water.	Casing is absolutely necessary, because the pressure at inlet to the turbine is much higher than the pressure at outlet. Unit has to be sealed from atmospheric pressure.
It is not essential that the wheel should run full and air has free access to the buckets.	Water completely fills the vane passage.

SOLVED PROBLEMS

Q.1- A power house is equipped with two water turbine each developing 2100 kW when running at 250 rpm under a maximum head of 23.4 m

a) Calculate the specific speed of the turbine.

b) State suitable type of the turbine and its runner

c) Determine approximately the inlet diameter of the runner if the speed coefficient Θ=0.8

d) Draw the typical inlet outlet triangles velocities.

Solution:-

$$P_t = 2100\ kW\ ; N = 250\ rpm\ ; H = 23.4\ m$$

a) $N_s = \frac{N\sqrt{P_t}}{H^{\frac{5}{4}}} = \frac{250\sqrt{2100}}{(23.4)^{\frac{5}{4}}} = 222.57\ m - kWunit$

b) $For\ N_s\ 50 \rightarrow 250\ Francis\ turbine\ is\ employed, Fast\ runner$

c) $Speed\ ratio\ \emptyset = 0.8$

$$\therefore\ U_1 = \emptyset\sqrt{2gH}\ and\ U_1 = \frac{\pi DN}{60}$$

$$\therefore\ D_1 = \frac{60 \times 0.8\sqrt{2 \times 9.81 \times 23.4}}{\pi \times 250} = 1.3\ m$$

d)

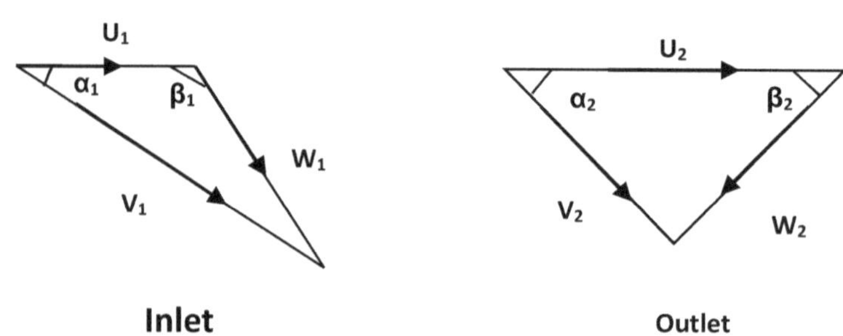

Q.2- A Kaplan turbine having an overall efficiency of 75% required to give 130 kW. The head is 6 m, velocity of periphery of runner is $0.8\sqrt{2gH}$ and the radial velocity of flow is $0.35\sqrt{2gH}$. the runner is to make 250 rpm and the hydraulic losses in the turbine 20% of the available energy. Determine

a) The angle of the guide blade at inlet.

b) The runner vane angle at inlet.

c) The diameter of the runner.

d) The width of the runner at inlet.

Assuming radial discharge $\alpha_2 = 90°$

Solution:-

$$P_t = 310\ kW\ ; N = 250\ rpm\ ; H = 6\ m$$

$$\eta_t = 0.95\ \ peripherel\ velocity$$

$$U_1 = 0.8\sqrt{2gH} = 0.8\sqrt{2 \times 9.81 \times 6}$$

$$\therefore U_1 = 8.68\ m/s$$

$$Velocity\ of\ flow\ V_{f1} = V_{f2} = V_2\ \ radial\ discharge$$

$$= 0.35\sqrt{2gH} = 0.35\sqrt{2 \times 9.81 \times 6} = 3.8\ m/s$$

$$losses = 0.2\ \text{Hydraulic energy}$$

$$As\ the\ available\ energy\ per\ unit\ weight\ of\ water = H\ m$$

$$and\ \eta_h = \frac{V_{u1}U_1}{gH} = 0.8$$

$$\therefore\ V_{u1} = \frac{9.81 \times 6 \times 0.8}{8.68} = 5.42\ m/s$$

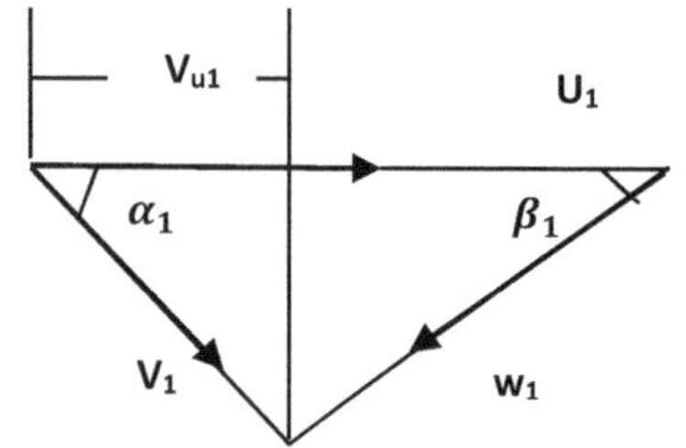

Inlet velocity triangle

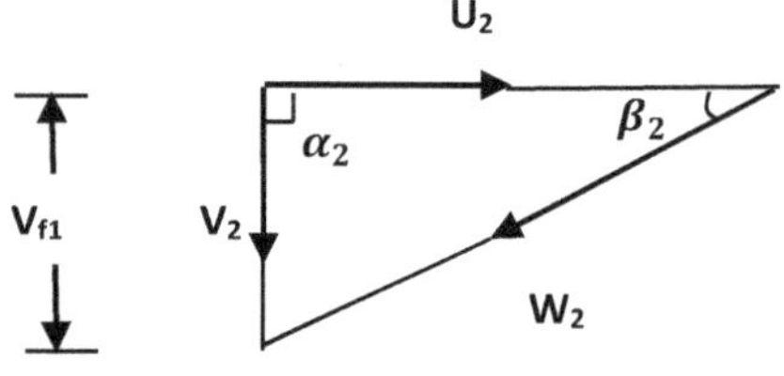

Outlet velocity triangle

a) angle of guide vane at inlet α_1

$$\alpha_1 = \tan^{-1}\frac{V_{f1}}{V_{u1}} = \tan^{-1}\frac{3.8}{5.42} = 35.03°$$

b) runner vane angle at inlet β_1

$$\tan\beta_1 = \frac{V_{f1}}{U_1 - V_{u1}} = \frac{3.8}{8.68 - 5.42} = 1.165$$

$$\therefore\ \beta_1 = 49.36°$$

c) $U_1 = \frac{\pi D_1 N}{60}$ $\quad or\ 8.68 = \frac{\pi \times D_1 \times 250}{60}$

$$D_1 = 0.663$$

d) with of the runner at inlet β_1

$$P_t = \gamma Q H \eta_t \qquad \therefore Q = \frac{130}{9.81 \times 6 \times 0.75} = 2.94\ m^3/s$$

$$and \quad Q = \pi D_1 B_1 V_{f1}$$

$$\therefore\ B_1 = \frac{2.94}{\pi \times 0.663 \times 3.8} = 0.371\ m$$

Q.3- A reaction turbine runner is required to operate under a head of 10 m at a speed of 175 rpm and to develop 175 kW. Find the diameter of the runner at inlet and outlet, the discharge, the guide vane angle and the runner vane angle at inlet and outlet, assuming the following data:

Outlet diameter = 0.66× inlet diameter

Discharge radial $\alpha_2 = 0\ ;\ U_1 = 0.75\sqrt{2gH}\ ;\ V_{f1} = 0.16\sqrt{2gH}$

$$;\eta_h = 0.86\ ;\ \eta_t = 0.81$$

Solution:-

$$P_t = 150\ kW\ ;\ N = 175\ rpm\ ; H = 10\ m$$

$$D_2 = 0.66D_1\ \ ;\ V_{f1} = V_{f2} = 0.16\sqrt{2gH}$$

$$U_1 = 0.75\sqrt{2gH}\ ;\ \eta_h = 0.86\ ;\ \eta_t = 0.81$$

a) diameter of the runner D_1

$$U_1 = 0.75\sqrt{2 \times 9.81 \times 10} = \frac{\pi D_1 N}{60} = 10.5\ m/s$$

$$and\ \ \therefore D_1 = \frac{60 \times 10.5}{\pi \times 175} = 1.146\ m$$

b) $D_2 = 0.66D_1 = 0.756\ m$

c) $Q = \frac{P_t}{\gamma H \eta_t} = \frac{150}{9.81 \times 10 \times 0.81} = 1.887\ m^3/s$

d) $\eta_h = \frac{V_{u1}U_1}{gH}\ \ for\ \alpha_2 = 90°$

$$\therefore\ V_{u1} = \frac{0.86 \times 9.81 \times 10}{10.5} = 8.03\ m/s$$

$$V_{f1} = 0.16\sqrt{2gH} = 0.16\sqrt{2 \times 9.81 \times 10} = 2.24\ m/s$$

e)

$$\tan\beta_1 = \frac{V_{f1}}{U_1 - V_{u1}} = \frac{2.24}{10.5 - 8.03}$$

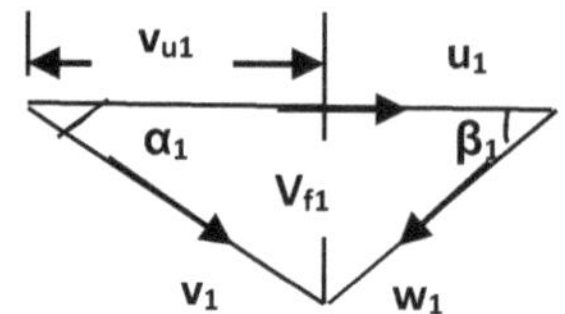

$$\therefore\ \beta_1 = 42.2°$$

f) Since

$$V_{f1} = V_{f2} = 2.24\, m/s$$

$$and\ U_2 = \frac{\pi D_2 N}{60} = \frac{\pi \times 0.756}{60} = 6.93\, m/s$$

$$\tan\beta_2 = \frac{V_{f1}}{U_2} = \frac{2.24}{6.93} = 0.323$$

$$\therefore\ \beta_2 = 179°$$

Q.4. A turbine works under a head of 20 m and makes 380 rpm. The outer and inner diameters are 60 cm and 30 cm respectively. The velocity of water entering the runner at 63 m/s and makes an angle of 10° with tangent. Assuming radial discharge. Determine the direction of the tangent to the vane of runner at inlet and outlet, also the hydraulic efficiency.

Solution:-

$$H = 20\, m\, ; N = 380\, rpm\, ;\ D_1 = 0.6\, m\, ;\ D_2 = 0.3$$
$$V_1 = 13\frac{m}{s}\, ;\ \alpha_1 = 10°\, ;\ V_f = constant\, ;\ \alpha_2 = 90°$$

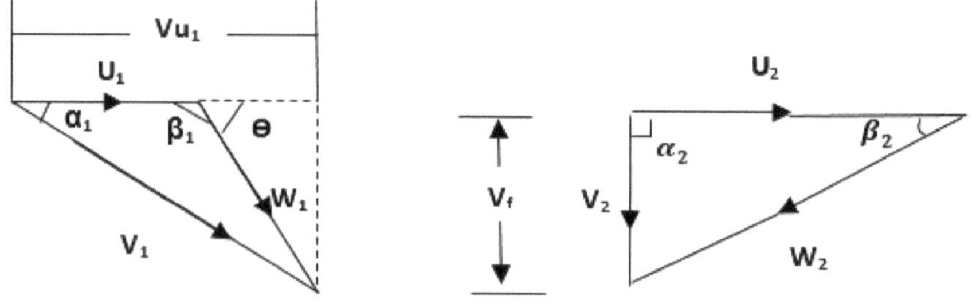

a) $V_{f1} = V_1 \sin \alpha_1$

$$= 13 \sin 10° = 2.26\, m/s$$

$$V_{u1} = V_1 \cos \alpha_1 = 13 \cos 10° = 12.8\, m/s$$

$$U_1 = \frac{\pi D_1 N}{60} = \frac{\pi \times 0.6 \times 380}{60} = 11.94\, m/s$$

$$\tan \theta = \frac{V_{f1}}{V_{u1} - U_1} = \frac{2.26}{12.8 - 11.94} = 2.628$$

$$\therefore\ \theta = 69.2°$$

b) $\tan \beta_2 = \frac{V_{f1}}{U_2}$ $\quad V_{f2} = V_{f1} = 2.628\, m/s$

$$U_2 = \frac{\pi D_2 N}{60} = \frac{U_1}{2} = 5.97\, m/s$$

$$\therefore\ \tan \beta_2 = \frac{2.628}{5.97}$$

$$\beta_2 = 20.73°$$

$$\eta_h = \frac{V_{u1} U_1}{gH} = \frac{12.8 \times 11.94}{9.81 \times 20} = 77.9\%$$

Q.5- Find the leading dimensions of the runner of a Francis turbine to develop 625 kW at 1000 rpm under a head of 100 m assuming a guide vane angle of 16°, axial length of the vane at inlet 0.1 times the outer diameter, inner radius 0.6 of outer radius, radial velocity of flow constant. Find discharge radial, hydraulic efficiency 0.88, overall efficiency 0.86, allowance thickness for vane thickness 5%.

Solution:-

$$P_t = 625\, kW\ ;\ N = 1000\, rpm\ ; H = 100\, m$$

$$D_2 = 0.6 D_1\ \ ;\ V_{f1} = V_{f2}\ ; K = 0.95$$

$$B_1 = 0.1 D_1\ ;\ \eta_h = 0.88\ ;\ \eta_t = 0.86$$

$$\alpha_1 = 16° ; \alpha_2 = 90°$$

$$\therefore P_t = \gamma QH\eta_t \; or \; 625 = 9.81 \times Q \times 100 \times 0.86$$

$$Q = 0.741 \frac{m^3}{s} = \pi D_1 B_1 V_{f1} K$$

$$= \pi D_1 \times 0.1 D_1 \times V_{f1} \times 0.95$$

$$V_{f1} = \frac{2.486}{{D_1}^2} \;\dots\dots\dots (1)$$

$$\eta_h = \frac{V_{u1} U_1}{gH} = 0.8 \; for \; \alpha_2 = 90° \dots\dots\dots (2)$$

$$and \; U_1 = \frac{\pi D_1 N}{60} = \frac{\pi D_1 \times 1000}{60} = 52.36 D_1 \; m/s$$

And substituting in (2)

$$0.8 = \frac{V_{u1} \times 52.36 D_1}{9.81 \times 100} \qquad \therefore V_{u1} = \frac{16.49}{D_1} \; m/s$$

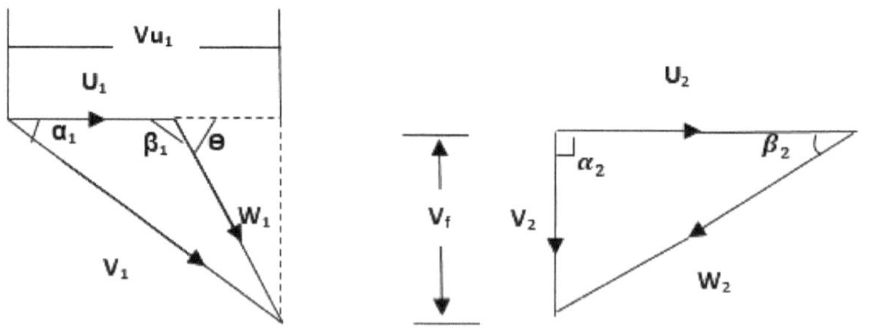

From inlet velocity triangle $\tan \alpha_1 = \frac{V_{f1}}{V_{u2}}$

$$\tan 16° = \frac{2.486}{{D_1}^2} \qquad \therefore D_1 = 0.526 \, m$$

$$\therefore D_2 = 0.6D_1 = 0.316\, m$$

And Brealth of runner at inlet

$$B_1 = 0.1D_1 = 0.0526\, m$$

Inlet

$$and \ \ \tan\theta = \frac{V_{f1}}{V_{u1} - U_1} \quad V_{u1} = \frac{16.49}{D_1} = \frac{16.49}{0.526}$$

$$\therefore V_{u1} = 31.35\, m/s$$

$$V_{f1} = \frac{2.486}{{D_1}^2} = 8.985\, m/s$$

$$U_1 = 52.36D_1 = 27.54\, m/s$$

$$\therefore \tan\theta = \frac{8.985}{31.35 - 27.54} = 2.358$$

$$\theta = 67°$$

$$\therefore \beta_1 = 180 - 67 = 113°$$

Outlet

$$\tan\beta_2 = \frac{V_{f2}}{U_2} \quad since\ V_{f1} = V_{f2} = 8.985\, m/s$$

$$U_2 = 0.6U_1 = 0.6 \times 27.54 = 16.52\, m/s$$

$$\therefore \ \beta_2 = \tan^{-1}\frac{8.985}{16.52} = 28.5°$$

$$\therefore \ Dimension\ \ D_1 = 0.526\, m \ \ D_2 = 0.316\, m$$

$$B_1 = 0.0526\, m \ \ \beta_1 = 113°\,; \ \beta_2 = 28.5°$$

Q.6- the following data refers to reaction turbine Q = 1 m³/s; H = 25 m; D_l =75 m; D_2 =50 m; V_{f2}=2.5 m/s; β_l = 35°, Calculate the power and rpm of the turbine. Assuming the width of the runner as constant and turbine efficiency 80%, hydraulic efficiency 82%.

Solution:-

$$Q = 1\,\frac{m^3}{s}\,; H = 25\,m\,;\; D_1 = 75\,cm\,; D_2 = 50\,cm$$

$$V_2 = 2.5\frac{m}{s}\,;\; \alpha_2 = 90°\,;\; \beta_1 = 35°$$

$$B_1 = B_2\,;\; \eta_h = 80\%\,;\; \eta_t = 82\%$$

a) $\eta_t = \frac{P_t}{\gamma QH}$ $\quad\therefore\; P_t = 0.8 \times 9.81 \times 25 \times 1 = 196.3\,kW$

b)

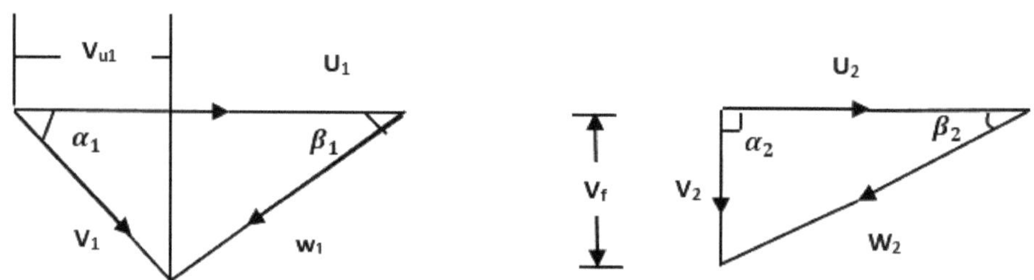

Inlet velocity triangle

Outlet velocity triangle

$$sine\; B_1 = B_2$$

$$and\;\therefore Q = \pi D_1 B_1 V_{f1} = \pi D_2 B_2 V_{f2}$$

$$\therefore\; D_1 V_{f1} = D_2 V_{f2}$$

$$and\; V_{f2} = V_2 \sin\alpha_2 = 2.5\,m/s$$

$$V_{f1} = \frac{0.5 \times 2.5}{0.75} = 1.67\,m/s$$

$$\eta_h = \frac{V_{u1}U_1}{gH}$$

$$\therefore V_{u1}U_1 = 0.82 \times 9.81 \times 25 = 201.11 \dots\dots\dots (1)$$

And from velocity triangle at inlet

$$\tan \beta_1 = \frac{V_{f1}}{U_1 - V_{u1}}$$

$$\therefore U_1 - V_{u1} = \frac{1.67}{\tan 35°} = 2.385 \dots\dots\dots (2)$$

From 1, 2

$$U_1 - \frac{201.11}{U_1} = 2.385$$

$$or\ {U_1}^2 - 2.385U_1 - 201.11 = 0$$

$$U_1 = \frac{-2.385 \pm \sqrt{(-2.385)^2 - 4(-201.11 \times 1)}}{2}$$

$$= \frac{-2.385 \pm \sqrt{5.69 + 884.44}}{2}$$

$$U_1 = 13.725 = \frac{\pi D_1 N}{60}$$

$$N = 349.5\ rpm$$

Q.7- A 233 l/s are supplied to a reaction turbine. The head available is 11 m. the wheel vanes are radial at inlet and the inlet diameter is twice the outlet diameter. The velocity of flow is constant and equal to 1.83 m/s. The runner makes 370 rpm Find:

a) guide vane angle

b) runner vane angle

c) inlet and outlet diameters of the runner. And

d) the width of the runner at inlet and outlet.

Neglect the thickness of the vanes (K=1). Assuming that the discharge is radial and that there are no losses in the runner.

Solution:-

$$Q = 0.233\ \frac{m^3}{s}\ ; H = 11\ m\ ;\ D_1 = 2D_2$$

$$\alpha_2 = 90°\ ;\ \beta_1 = 90°$$

$$V_{f1} = V_{f2} = 1.83\frac{m}{s}\ ; N = 370\ rpm$$

$$No\ lossos\ in\ runner.\ V_{f1} = V_{f2}$$

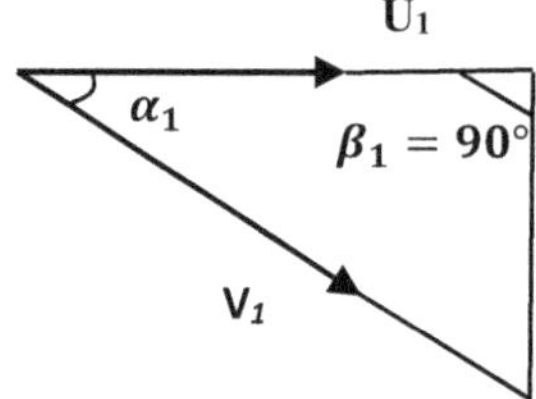

Inlet velocity triangle

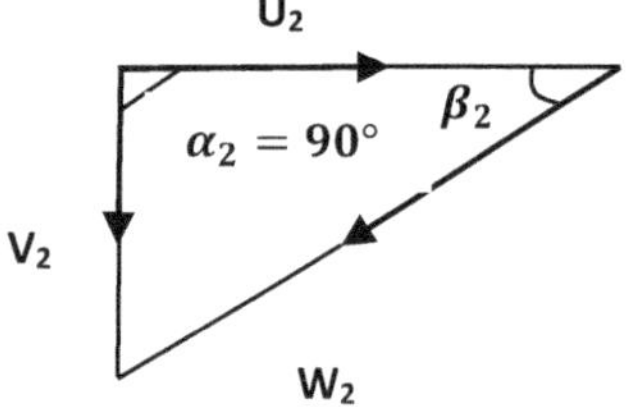

Outlet velocity triangle

a) Work done by the runner per unit weight

$$\frac{V_{u1}U_1}{g} = H - \frac{{V_2}^2}{2g} \qquad V_2 = V_{f2}$$

$$\therefore\ \frac{{U_1}^2}{g} 11 - \frac{(1.83)^2}{2g}$$

$$U_1 = 10.3\ m/s$$

Now

$$\tan\alpha_1 = \frac{V_{f1}}{U_1} = \frac{1.83}{10.3}$$

$$\alpha_1 = 10°$$

b) $since\ \beta_1 = 90°\ radial\ at\ inlet$

c) $U_1 = \frac{\pi D_1 N}{60} \quad \therefore\ D_1 = \frac{10.3\times 60}{\pi\times 370} = 0.532\ m$

d) $D_2 = \frac{1}{D_1} = 0.266\ m$

e) $Q = \pi D_1 B_1 V_f$

$$0.233 = \pi \times 0.532 \times B_1 \times 1.83$$

$$B_1 = 0.0762\ m = \frac{1}{2}\ B_2$$

$$or\ B_2 = 2B_1 = 0.1524\ m$$

Q.8- A turbine runner has an exit velocity of 10 m/s. the loss of head due to friction and other causes in the draft tube should not exceed 1.5 m. what maximum height of setting will you recommend for the turbine if the cavitation is to be avoided?

Assuming:

i) The velocity of water at the outlet of draft tube as 2.5 m/s.

ii) The cavitation commences when the pressure is 2.5 m of water absolute.

Solution:-

$$V_2 = 10\frac{m}{s}\ ;\ H_{L_{2-3}} = 1.5\ m\ ;\ V_3 = 2.5\frac{m}{s}$$

$$\frac{P_V}{\gamma} = 2.5\ m\ of\ water$$

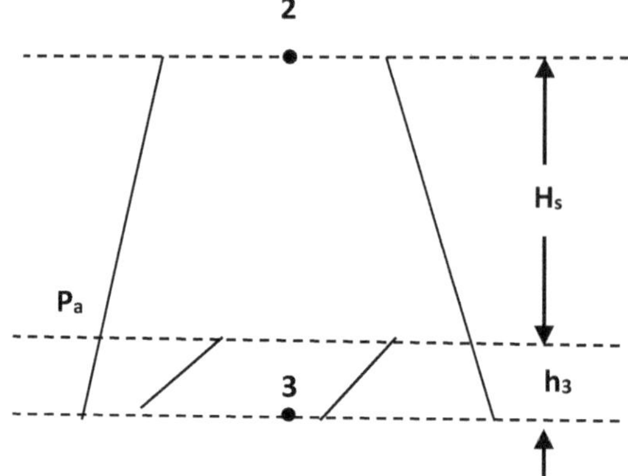

Apply B.E between 2, 3

$$\frac{P_2}{\gamma} + \frac{V_2^{\,2}}{2g} + Z_2 = \frac{P_3}{\gamma} + \frac{V_3^{\,2}}{2g} + Z_3 + H_{L_{2-3}}$$

$$\frac{P_2}{\gamma} = \frac{P_3}{\gamma} - (Z_2 - Z_3) - \frac{(V_2^{\,2} - V_3^{\,2})}{2g} + H_{L_{2-3}}$$

$$\frac{P_3}{\gamma} = \frac{P_a}{\gamma} + h_3\ \ and\ Z_2 - Z_3 - h_3 = H_s$$

$$\frac{P_2}{\gamma} = \frac{P_a}{\gamma} - H_s - \frac{(10^2 - 2.5^2)}{2g} + 1.5\ \ since\ \frac{P_2}{\gamma} = \frac{P_V}{\gamma}$$

$$2.5 = 10.3 - H_s - 4.778 + 1.5$$

$$H_s = 4.52\ m$$

Q.9- The following data were obtained from an efficiency test of Kaplan turbine:-

Diameter of the boss of runner = 0.35 times external diameter

Speed ratio Ø =2 flow ratio Ψ =0.6

Power output = 13250 kW Head = 8 m

Find the efficiency of the turbine.

Solution:-

$$D_2 = 0.35D_1\,;\; \emptyset = 2\,;\; \varphi = 0.6$$

$$N = 75\,rpm\,; H = 8\,m\,:\; P_t = 13250\;kW$$

$$since\;V_f = 0.6\sqrt{2gH} = 0.6\sqrt{2\times 9.81\times 8}$$

$$= 7.5\;m/s$$

$$U_1 = \emptyset\sqrt{2gH} = 2\sqrt{2\times 9.81\times 8}$$

$$= 25.06\;m/s$$

$$U_1 = \frac{\pi D_1 N}{60} \quad \therefore\; D_1 = \frac{25.06\times 60}{\pi\times 75}$$

$$D_1 = 6.38\;m$$

$$\therefore\; Q = \frac{\pi}{4}\left({D_1}^2 - {D_2}^2\right)V_f$$

$$= \frac{\pi}{4}\left((6.38)^2 - (0.32\times 6.38)^2\right)\times 7.5$$

$$= 210\;m^3/s$$

$$\eta_t = \frac{P_t}{\gamma QH} = \frac{13250}{2\times 9.81\times 8} = 80.3\%$$

Q.10- Kaplan turbine installed at power house develops 1300 kW at average head of 29 m. the speed and flow ratios are 2.1 and 0.62 respectively. d = 0.35 D_1. Overall efficiency 0.89, calculate the diameter and the speed of the runner.

Solution:-

$$D_2 = 0.34D_1\,;\; \emptyset = 2.1\,;\; \varphi = 0.62$$

$$\eta_t = 0.81\ ;\ H = 29\ m\ :\ P_t = 1300\ kW$$

$$\therefore Q = \frac{P_t}{\gamma H \eta_t} = \frac{1300}{9.81 \times 29 \times 0.81} = 5.13\ m^3/s$$

$$and\ V_f = 0.62\sqrt{2gH} = 0.6\sqrt{2 \times 9.81 \times 29}$$

$$= 14.787\ m/s$$

$$and\ Q = \frac{\pi}{4}\left(D_1{}^2 - D_2{}^2\right)V_f$$

$$5.13 = \frac{\pi}{4}\left(D_1{}^2 - (0.34D_1)^2\right) \times 14.787$$

$$D = 0.707\ m$$

$$U_1 = \emptyset\sqrt{2gH} = 2.1\sqrt{2 \times 9.81 \times 29}$$

$$= 50\ m/s$$

$$U_1 = \frac{\pi D_1 N}{60} \quad \therefore\ N = \frac{50 \times 60}{\pi \times 0.707} = 1350\ rpm$$

Q.11- Each Kaplan turbine at Vargon (Sweden) in rated at 11350 kW when working under an average head of 4.3 m. the diameter of the hub is 0.3 times the external diameter of runner. Overall efficiency of turbine is 0.91. Find the speed and diameter of runner. Take the values of Ø = 2 and Ψ = 0.65 respectively.

Solution:-

$$D_2 = 0.3D_1\ ;\ \emptyset = 2\ ;\ \varphi = 0.65$$

$$\eta_t = 0.91\ ;\ H = 4.3\ m\ :\ P_t = 11250\ kW$$

a) $$Q = \frac{P_t}{\gamma H \eta_t} = \frac{11250}{9.81 \times 4.3 \times 0.91} = 296\ m^3/s$$

$$V_f = \varphi\sqrt{2gH} = 0.65\sqrt{2 \times 9.81 \times 4.3}$$

$$= 5.97\ m/s$$

$$and\ Q = \frac{\pi}{4}\left(D_1{}^2 - D_2{}^2\right)V_f$$

$$296 = \frac{\pi}{4}\left(D_1{}^2 - (0.3D_1)^2\right) \times V_f$$

$$D_1 = 8.33\ m$$

b) $U_1 = \emptyset\sqrt{2gH} = U_1 = \frac{\pi D_1 N}{60}$

$$= 2\sqrt{2 \times 9.81 \times 4.3} = \frac{\pi \times 8.33N}{60}$$

$$\therefore\ N = 42.1\ rpm$$

CHAPTER 4

Similarity Laws for Turbine Specific Speed and Cavitations

Abstract: It is possible that a hydraulic machine will not give the desired result for which it has been designed. Such a machine is costly for manufacture and once it is made, it is difficult to change its components. Therefore, it is required to predict the performance of a prototype hydraulic machine before it is manufactured. This is done by making its model. Experiments are first performed on models from their results the performance of the prototype machine is predicted chapter on "Dimensional and Model Analysis" given in Fluid Mechanics will be useful for the prototype. Here prototype and its model are two similar machines having different specifications, which are to be compared. The concept of "Unit and specific Quantities" is a prerequisite for comparison of hydraulic machines. In addition, in this chapter a brief discussion about cavitation in turbine, turbine section, marking types of turbine and hydraulic turbine classification and section according to hydraulic power plants.

Keywords: Cavitation in turbines, Similarity laws, Turbines classification and selection.

4.1. SIMILARITY LAWS

Two similar hydraulic machines designed for different specifications are required to compare.

It is possible that a hydraulic machine may not give the desired results for which it has been designed. Such a machine is costly to manufacture and once it is made, it is difficult to change its components. Therefore, it is required to predict the performance of a prototype hydraulic machine before it is manufactured. This is done by making its model. The rate of flow, speed, power, *etc.* of hydraulic machines all functions of the working head which is one of the most fundamental of all quantities that go to determine the flow phenomena associated with machines such as turbines and pumps. For this reason, specific quantities are obtained by reducing any quantity to a corresponding unit head and some size such as the diameter of the runner for a reaction turbines and least jet diameter of Pelton turbines [1, 2].

Jafar Mehdi Hassan, Salman Hussien Omran, Laith Jaafer Habeeb,
Alamaslamani Ammar Fadhil Shnawa & Adrian Ciocănea

1- Specific Speed: it is the speed of identical turbine (geometrically similar and having same blade angle) working under the unit head and delivery unit power.

The concept of specific is very important in the study of turbines and pumps. It is a modern basis of scientific classification of turbines and pumps.

It has been shown in section (3.2) that the tangential velocity at the runner is

$$U_1 = \frac{\pi D_1 N}{60} \quad or\ U_1 \propto D_1 N$$

And

$$U_1 = \phi\sqrt{2gH} \ \ or\ U_1 \propto \sqrt{H}$$

$$\therefore \ \ ND_1 \propto \sqrt{H} \ \rightarrow \ D_1 \propto \frac{\sqrt{H}}{N} \qquad \textbf{(4.1)}$$

Also, the power developed by the turbine is

$$P_t = \gamma Q H_n \quad or \ \ P_t \, \alpha \, QH_n$$

And

$$Q = AV \quad and\ V = \varphi\sqrt{2gH}$$

$$\therefore \ \ Q \propto {D_1}^2 \sqrt{H}$$

$$or\ \ P_t \, \alpha \, {D_1}^2 \sqrt{H}.H$$

$$or\ \ P_t \, \alpha \, {D_1}^2 H^{3/2} \qquad \textbf{(4.2)}$$

From (**4.1, 4.2**) substituting for D_1 we get

$$P_t \, \alpha \frac{H}{N^2} H^{\frac{3}{2}} \quad i.e\ \ P_t \, \alpha \frac{H^{\frac{5}{2}}}{N^2}$$

$$or\ N \propto \sqrt{\frac{H^{\frac{5}{2}}}{P_t}} \quad or\ N = N_s\sqrt{\frac{H^{\frac{5}{2}}}{P_t}}$$

$$\therefore N_s = \frac{N\sqrt{P_t}}{H^{\frac{5}{4}}} \qquad \textbf{(4.3)}$$

$If\ \ P_t = 1\ kW\ and\ H = 1\ m\ then\ numerically$

$N_s = N\ Which\ called\ unit\ speed$

$P_t = power\ developed\ by\ the\ turbine\ kW$

$N = speed\ of\ the\ runner\ in\ rpm$

$H = \ net\ head\ \ m$

During the design of the turbine, the specific speed should be the same for the model and prototype, therefore:

$N_{sm} = N_{sp}$

$m = model\ \ ;p = prototype$

$\therefore\ from\ equ. 3,2\ for\ P_{tm}\ \&\ P_{tp}$

$$\frac{N_m}{N_p} = \frac{D_p}{D_m} \times \sqrt{\frac{H_m}{H_p}} \qquad \textbf{(4.4)}$$

2- Specific Flow:

For reaction turbine

$$Q = \pi DBV_f$$

The dimensions D and B generally have linear relations with D_1 the runner diameter at the inlet and therefore, since

$$V_f \propto \sqrt{H}$$

$$Q \propto {D_1}^2\sqrt{H}$$

$$or\ Q = Q_s\,{D_1}^2\sqrt{H}$$

$$\therefore\ Q_s = \frac{Q}{{D_1}^2\sqrt{H}} \qquad \textbf{(4.5)}$$

$$D_1 = the\ outlet\ diameter\ of\ the\ runner\ m$$

$$H = net\ head\ \ m$$

$$Q = flow\ rate\ m^3/s$$

Also for Pelton turbine

$$Q_s = \frac{Q}{{d_1}^2\sqrt{H}} \qquad \textbf{(4.6)}$$

$$d_1 = diameter\ of\ the\ jet\ (m)$$

Then for similarity low

$$Q_{sm} = Q_{sp}$$

$$or\ \ \frac{Q_m}{Q_p} = \left(\frac{D_m}{D_p}\right)^2 \times \sqrt{\frac{H_m}{H_p}} \qquad \textbf{(4.7)}$$

3- Specific Power:

$$Since\ P_t = \gamma QH$$

Therefore, for reaction turbine

$$P_{ts} = \frac{P_t}{{D_1}^2 H^{3/2}} \qquad \textbf{(4.8)}$$

And for Pelton turbine

$$P_{ts} = \frac{P_t}{{d_1}^2 H^{3/2}}$$

And for similarity low

$$\frac{P_m}{P_p} = \left(\frac{D_m}{D_p}\right)^2 \times \left(\frac{H_m}{H_p}\right)^{3/2} \quad \textbf{(4.9)}$$

Since the power P=Tw where T is the torque, therefore

$$\frac{T_m}{T_p} = \left(\frac{D_m}{D_p}\right)^2 \times \frac{H_m}{H_p} \quad \textbf{(4.10)}$$

$$Where\ w \propto \sqrt{H}$$

The scale ratio or model ratio: it is the ratio of the runner diameters of the prototype to the diameter of the model *i.e.*

$$n = \frac{D_p}{D_m}\ or\ \frac{d_p}{d_m}$$

4.2. CAVITATION IN TURBINES

- Cavitation is a term used to describe a process that includes nucleation growth and the implosion of vapor or gas-filled cavities. These cavities are formed into a liquid when the static pressure of the liquid for one reason or another is reduced below its vapor presser at the prevailing temperature. When cavities are caved to the high-pressure region, they implode violently.

- Cavitation is an undesirable effect that results in pitting, mechanical vibration and loss of efficiency.

- If the nozzle and buckets are not properly shaped in impulse turbine, flow separation from the boundaries may occur at same operating conditions that may cause regions of low pressure and result in cavitation.

- The turbine parts exposed to cavitation are the runner, draft tube cones for the Francis and Kaplan turbine and the needles, nozzle and the runner buckets of the Pelton turbines [16].

- Measures for compacting and damage under cavitation conditions include improvements in hydraulic design and production of components with erosion resistant materials and arrangement of the turbine for operations within good range of acceptable cavitation conditions.

(Fig. **4.1, a-c**) show traveling bubble cavitations and damages in turbines

From the draft tube theory, it is found that the cavitation depends on the following:-

a) Vapor pressure, H_v which is a function of temperature of flowing water.

b) Absolute pressure, $(H_a - H_v)$ or barometric pressure due to the location of turbine above mean sea level.

c) Suction pressure, H_s which is the high of the runner outlet above the tailrace level.

d) Effective dynamic suction head and absolute of water at runner exit and V_2 respectively.

The dynamic head = $\eta_d \left(\frac{V_2{}^2 - V_3{}^2}{2g}\right)$referring to draft tube.

According to Prof. Thoms cavitation can be avoided if the values of σ are not less than the critical value given by the manufacturer. *i.e.*

$$\sigma \leq \sigma_{critiol}$$

$$\sigma_{critiol} = \frac{\Delta H_t}{H} = \frac{\text{Theoretical head of the turbine}}{Working\ head\ of\ turbine\ (m)}$$

$$\Delta H_t = \frac{\Delta P}{\gamma} + \frac{V_2{}^2}{2g} + Z_2 + losses$$

Then the cavitation empirical formula as follows for Francis turbine

$$\sigma = 0.625 \left(\frac{N_s}{100}\right)^2 \quad \textbf{(4.11)}$$

Kaplan turbine

$$\sigma = 0.28 + \frac{1}{7.5}\left(\frac{N_S}{100}\right)^2 \qquad \textbf{(4.12)}$$

General

$$\sigma = \frac{H_a - H_v - H_s}{H_n} \qquad \textbf{(4.13)}$$

Fig. (**4.2**) show the relation between cavitation parameter and specific speed for the reaction turbines.

(**a**)
Traveling bubble cavitation in Francis turbine

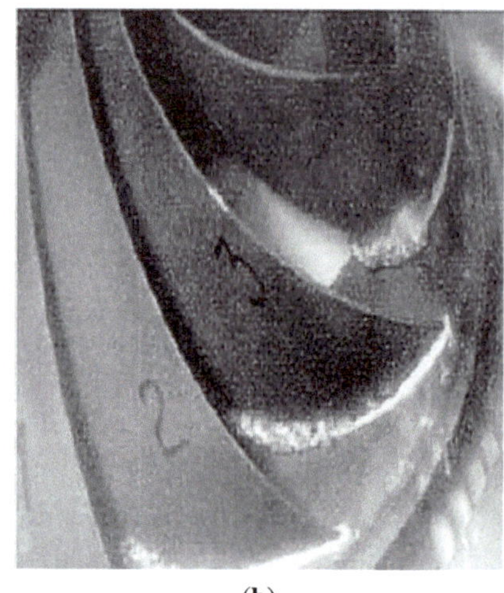

(**b**)
Inlet edge cavitation in Francis turbine

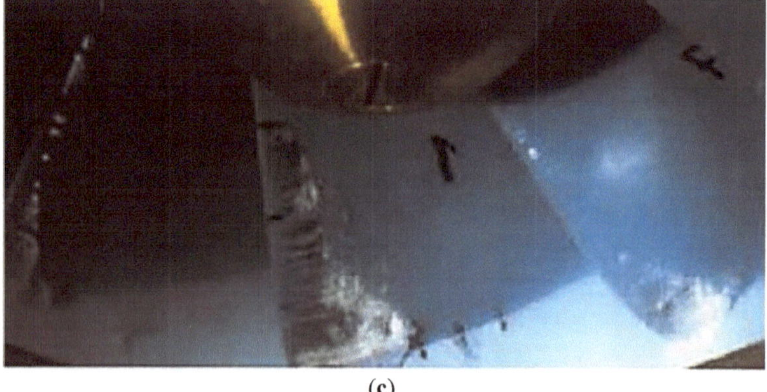

(**c**)
Leading edge cavitation damage in Francis turbine

Fig. (4.1). Cavitation in Turbine [7].

- The value of σ, at which cavitation will occur, is the critical value.
- Typical values of the critical cavitation parameter for reaction turbine are shown.

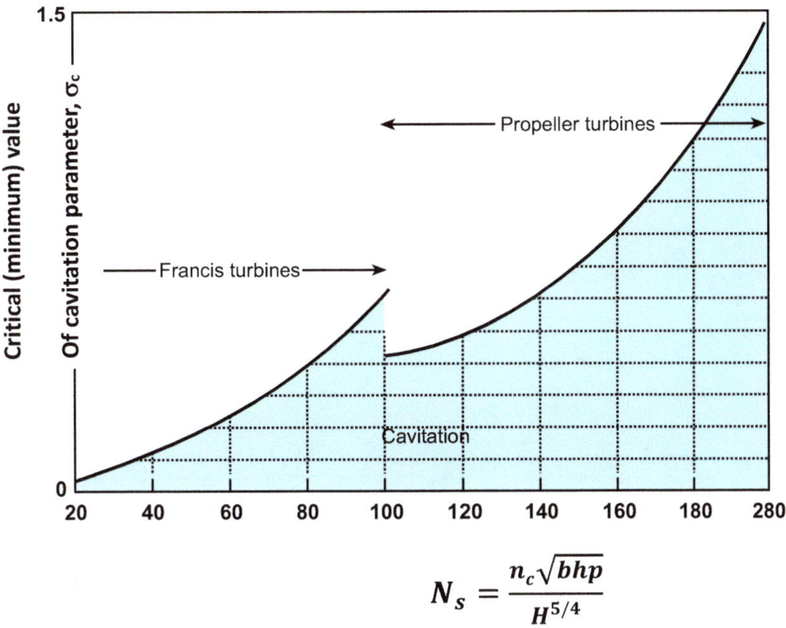

$$N_s = \frac{n_c\sqrt{bhp}}{H^{5/4}}$$

Fig. (4.2). Critical Value of Cavitation Parameter [7].

EX: A quarter scale turbine model is tested under a head of 36 m. the full-scale turbine is required to work under a head of 100 m and to run at 428 rpm. At what speed must the model be run and if it develops 100 kW and uses 0.324 m^3/s of water at this speed. What power will be admitted from the fall scale turbine assuming that its efficiency is 3% better than that of the model?

Solution:-

$$H_m = 36\,m\ ;\ H_p = 100\,m\ ;\ N_p = 428\,rpm$$

$$Q_m = 0.324\,m^3/s\ ;\ P_t = 100\,kW$$

$$scale\ ratio\ \ m = \frac{D_p}{D_m} = 4$$

$$Q_{sm} = Q_{sp}$$

$$\frac{Q_m}{{D_m}^2\sqrt{H_m}} = \frac{Q_p}{{D_p}^2\sqrt{H_p}}$$

$$or \quad Q_p = Q_m\left(\frac{D_p}{D_m}\right)^2\left(\frac{H_p}{H_m}\right)^2$$

$$Q_p = 0.324(4)^2\left(\frac{100}{36}\right)^{\frac{1}{2}} = 8.65\frac{m^3}{s}$$

$$P_{tp} = \gamma Q_p H_p \eta_p$$

$$= 9.81 \times 8.65 \times 100 \times \eta_p$$

$$Since \ \eta_m = \frac{P_{tm}}{\gamma Q_m H_m} = \frac{100}{9.81 \times 0.324 \times 36} = 0.87$$

$$\eta_p = 0.87 + 0.03 = 0.9$$

$$\therefore \ P_{tp} = 7637\ kW$$

$$N_{sp} = N_{sm}$$

$$\frac{N_p\sqrt{P_{tp}}}{{H_p}^{\frac{5}{4}}} = \frac{N_m\sqrt{P_{tm}}}{{H_m}^{\frac{5}{4}}}$$

$$N_m = N_p \times \left(\frac{P_{tp}}{P_{tm}}\right)^{\frac{1}{2}} \times \left(\frac{H_m}{H_p}\right)^{\frac{5}{4}}$$

$$= 428 \times \left(\frac{7637}{100}\right)^{\frac{1}{2}} \times \left(\frac{36}{100}\right)^{\frac{5}{4}}$$

$$= 1027\ rpm$$

$$\therefore\ N_{sm} = 116.44$$

EX: A 11.77 MW reaction water turbine has effective head of 25 m. The runner is 3 m above the tail race. The turbine is installed at an elevation of 300 m above sea level when the barometric head is 9.64 m of water. The temperature of water is 27°C. The following table give values if the plant sigma σ (Thoma's cavitation factor) against specific speed the turbine is coupled to a 50 cycles alternator. Calculate the synchronous speed of the turbine?

σ	0.05	0.10	0.16	0.228
N_s	116	155	193	230

Solution:-

$$P_t = 11.77\ kW\ ;\ H_n = 25\ m\ ;\ H_s = 3\ m\ ;\ H_b = 9.64$$

$$\therefore\ \sigma = \frac{H_b - H_s}{H_n} = \frac{9.64 - 3}{25} = 0.2646$$

The value of specific speed N_s corresponding to σ is found from the following curve drawn with help of given table. Thus $N_s \leqq 240$. It is clear from the curve that if $N_s > 240$, then cavitation appear.

$$Now\ N_s = \frac{N\sqrt{P_t}}{H^{\frac{5}{4}}} = \frac{N\sqrt{11770}}{(25)^{\frac{5}{4}}} = 240$$

$$or\ N = 123.5\ rpm$$

Assuming P, the number of pair poles for the alternator as 24, the synchronous speed can be found with help of

$$f = \frac{P.N}{60}$$

$$or\ 50 \times 60 = 24 \times N$$

$$\therefore\ N = 125\ rpm$$

Which is the nearest synchronous speed to 123.5 rpm.

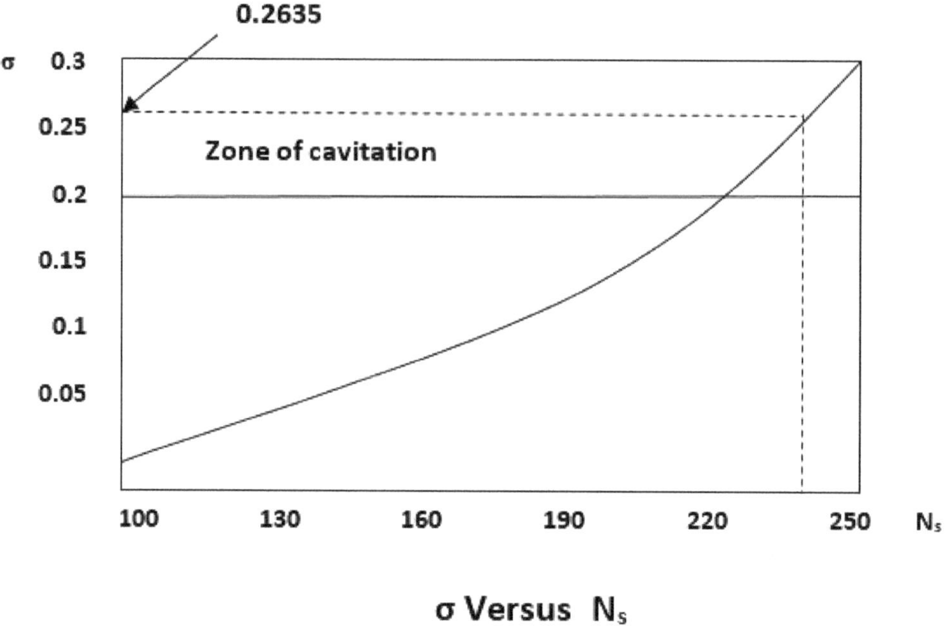

σ Versus N_s

EX.3: In a hydroelectric station water is available at the rate of 175 m^3/s under a head of 18 m. If the available turbine run at a speed of 150 rpm. With overall efficiency of 82%. Find the minimum number of turbine of the same size required in case of

(i) Francis turbine with max specific speed of 260
(ii) Kaplan turbine of N_s =350

Solution:-

$$Q = 175\frac{m^3}{s} \ ; H_n = 18\,m \ ; N = 150 \ ; \eta_0 = 0.82$$

$$\therefore \eta_0 = \frac{P_{out}}{P_{in}} = \frac{P_t}{\gamma Q H_n}$$

$$P_t = 0.82 \times 9.81 \times 175 \times 18 = 25339\,kW$$

i) No. of Francis turbine

$$\therefore\ N_s = \frac{N\sqrt{P_t}}{H^{\frac{5}{4}}} = 260 = \frac{150\sqrt{P}}{(18)^{\frac{5}{4}}}$$

$$\therefore\ P_t = 4129\ kW$$

$$\therefore\ No.of\ turbine = \frac{25339}{4129} = 6.135 \approx 7$$

Also for Kaplan turbine

$$N_s = 350 = \frac{150\sqrt{P}}{(18)^{\frac{5}{4}}}$$

$$\therefore\ P_t = 7484\ kW$$

$$\therefore\ No.of\ turbine = \frac{25339}{7484} = 3.38 \approx 4$$

4.3. TURBINE SELECTION

The selection of turbine is an essential stage in designing a HEPP when operating conditions of the turbine (head, power) must be taken into account, especially the layout of the power house, the supply and tail water conduits, construction and service conditions, engineering and cost data [10].

Some turbine elements are of such great size, that almost always they prescribe the dimension and layout of the power house of HEPP. The interrelation between the turbine and the building structure acquires a greater importance.

A HEPP unit consists of two machines- a turbine and a generator, they have a common system of bearings.

The turbine runner and generator rotor are rigidly connected by common shaft, the rotate with an equal speed.

The synchronous speed n_{syn} is determined by two formulae

For 50 H_z n_{syn}=6000/P

For 60 H_z n_{syn}=7200/P

Where P is the number of generator poles that must be even, and at P>24 it is desirable to be multiple to four. Thus, a number of n_s values is defined and intermediate values are impossible.

4.4. MARKING TYPES OF TURBINE

The nomenclatures permit the selection of turbines when designing a HEPP. Fig. (**4.3**) shows the nomenclature (range chart) of large adjustable blade (axial flow) and radial turbine. It is clear that the nomenclature corresponds to heads ranging from 3-4 to 500 m, and powers from 1-2 MW to 300-800 MW. Fig (**4.4**) demonstrates the previously used nomenclature of small turbine corresponding to heads ranging from 1.5 to 100 m and powers within 10-2500 MW.

An abbreviated marking of turbine is often used with the mark including four indices.

1. The kind of a turbine is denoted by letters:

II JT - adjustable – blade axial turbine (AB in Fig. (**4.3**))

II or II JT TS - adjustable – blade mixed-flow turbine (ABM)

II JT K - adjustable – blade bulb-type turbine (aBB)

PO - radial-axial turbine (RA in Fig. (**4.3**))

II P - Propeller-type wheel axial turbine (PA)

II P A - Propeller-type mixed-flow turbine (PM)

K - Pelton turbine (P)

2. The type of turbine is mainly determined by the head. Several types of turbine of the same head are available, differing only by the form of the water passage. Each type of turbine is assigned a serial No. that is given.

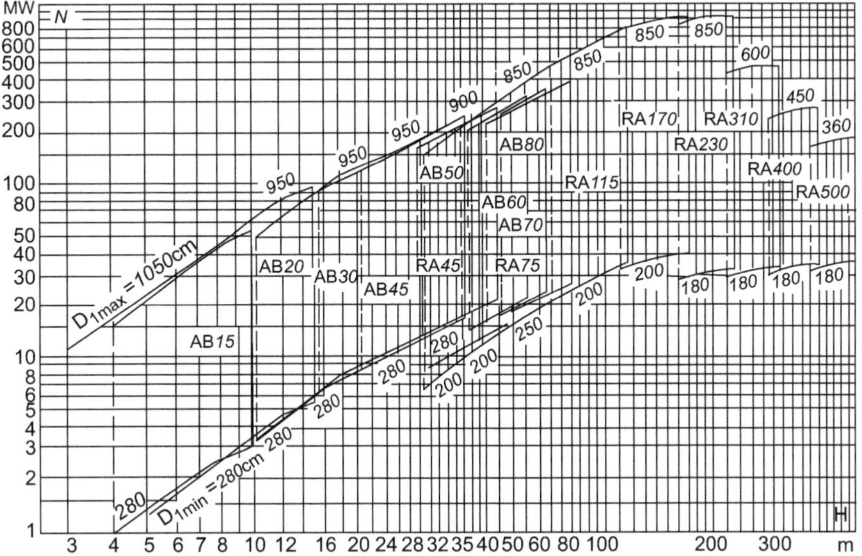

Fig. (4.3). Nomenclature of axial-flow adjustable – blade and radial-axial flow turbines [6].

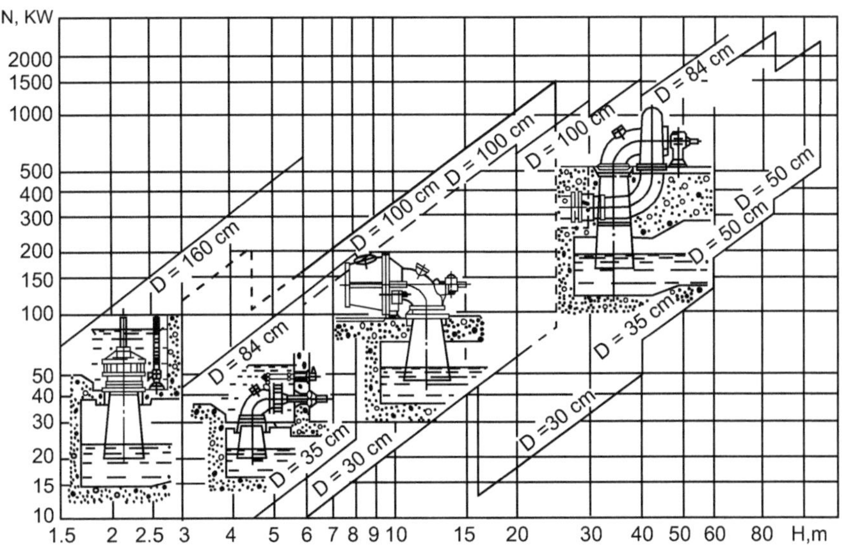

Fig. (4.4). Nomenclature of small turbines [6].

In the turbine marking (sometimes as a fraction: the numerator is the maximum head, and the denominator the type of turbine.

3. The layout is determined by the position of the shaft of the turbine-generator set and it can be vertical B or horizontal Γ.

4. Nominal turbine diameter D_1 (Figs. **4.5** and **4.6**). For diagonal turbine the runner blade angle Θ and for Pelton turbine the nozzle diameter and the number of jets are given.

Example of turbine markings:

II JT 20/811-B-800	is the marking of an axial-flow adjustable – blade turbine, maximum head is 20 m, type of water passage No. 811, B-vertical D_1 =8 m.
II 120/45-2556-B-600	mixed flow adjustable – blade Turbine, maximum head is 120 m, Θ=45° type No. 2556, B-vertical, nominal dia. D_1 = 6 m.
II JT K 15/548- -600	is the marking of an axial-flow adjustable – blade Bulb-type turbine, maximum head is 15 m, the type of water Passage No. 548, - horizontal Γ diameter = 6 m.

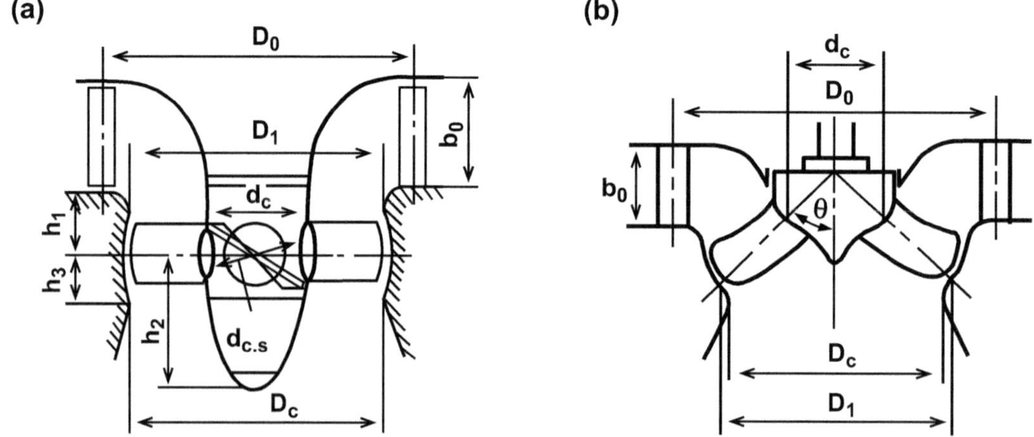

Fig. (4.5). Basic dimensions of adjustable- turbine [6]: (**a**) axial- flow; (**b**) mixed-flow.

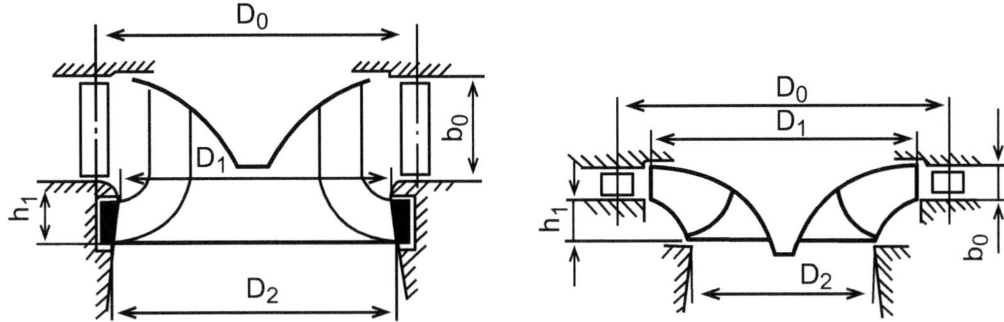

Fig. (4.6). Basic dimensions of radial –axial flow turbines [6].

4.5. HYDRAULIC TURBINES CLASSIFICATION AND SELECTION

The following table gives an overview of reference values of specific speed and stage reaction for different hydraulic turbines.

The specific speed increases as flow rate increases and hydraulic head decreases.

Therefore, turbines with high specific speed have also high values of stage reaction, because work exchange between fluid and runner decreases if R increases. Where R is the ratio of piezometric head charge in the runner and draft tube and the total piezometric head charge.

	K	N_s	R
Pelton 1 jet	0.05 ÷ 0.2	5 ÷ 10	0
Pelton 2 jets	0.1 ÷ 0.3	7 ÷ 14	0
Pelton(>2 jets)	0.3 ÷ 0.4	14 ÷ 20	0
Francis ("slow")	0.3 ÷ 0.6	15 ÷ 33	0.30
Francis ("medium")	0.6 ÷ 1.0	33 ÷ 55	0.40
Francis ("fast")	1.0 ÷ 1.6	55 ÷ 80	0.50
Francis ("ultrafast")	1.6 ÷ 2.3	80 ÷ 120	0.60
Propeller, Kaplan	1.4 ÷ 5.7	75 ÷ 300	0.70

Where

$$K = \gamma \frac{Q^{1/2}}{(gH)^{3/4}} \qquad N_s = N \frac{\sqrt{P_t}}{H^{5/4}} \qquad R = \frac{\Delta H_{p.r}}{\Delta H_p}$$

The value of working parameters and output for the main types of hydraulic turbines (Fig. **4.7**) shown as follows.

- Pelton:

  Flow rate ~ 0.5 ÷20 m^3/s

  Head ~ 300 ÷1500 m

  Net power up to ~ 200 MW

- Francis:

  Flow rate ~ 2 ÷800 m^3/s

  Head ~ 500 ÷400 m

  Net power up to ~ 800 MW

- Kaplan:

  Flow rate ~ 1000 ÷20 m^3/s

  Head ~ 40 m

  Net power up to ~ 200 MW

High head power plant

turbina Pelton

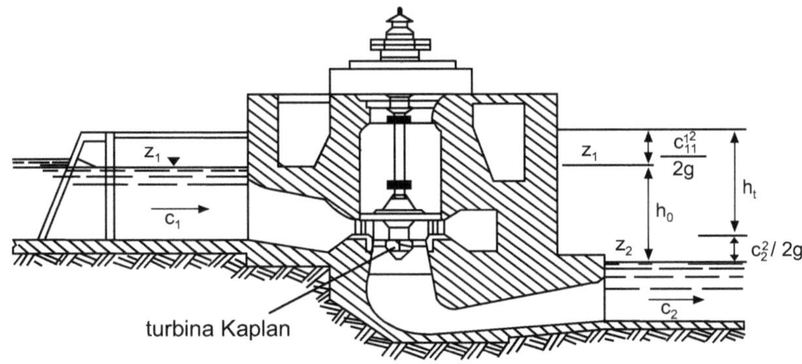

Low head power plant

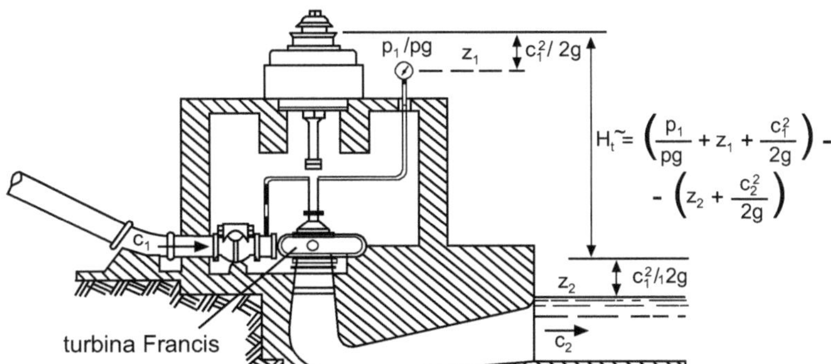

Medium head power plant

Fig. (4.7). Hydraulic Turbine Classification [6].

The relation between the net head of a turbines and the power generated show in Fig. (**4.8**).

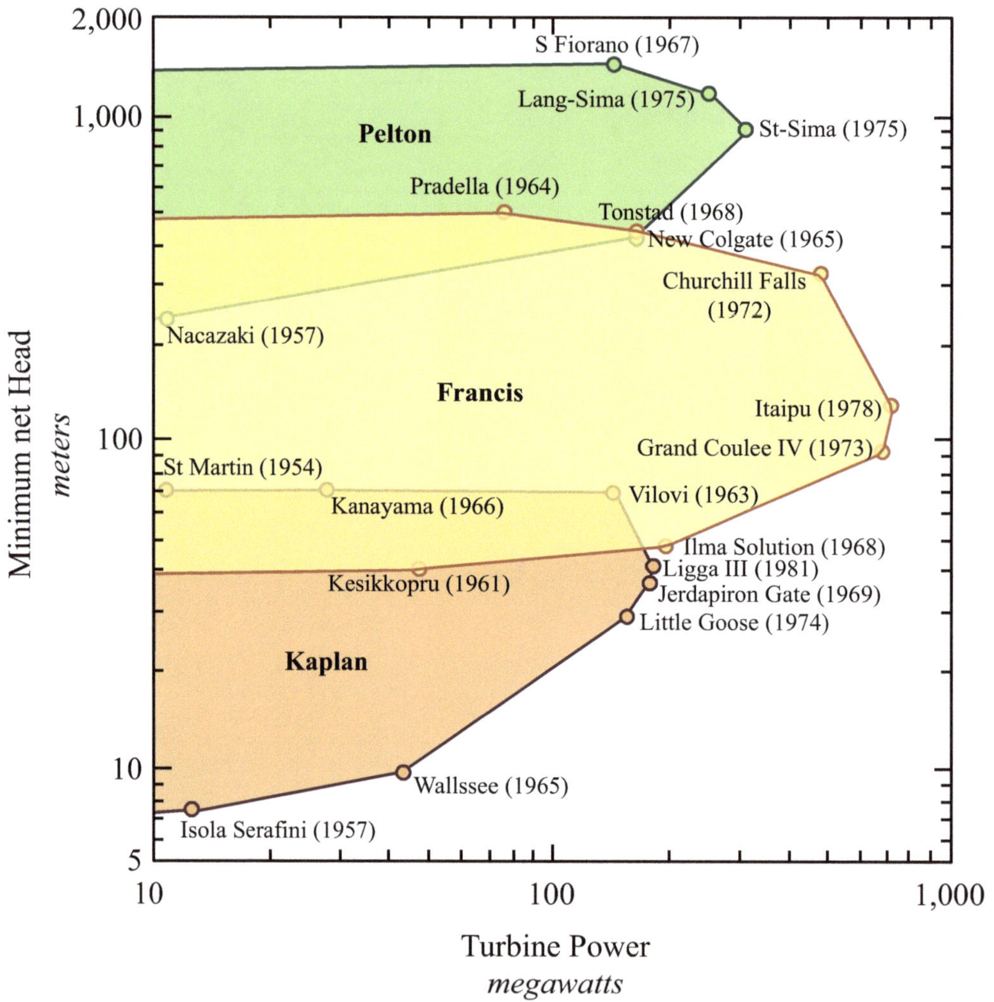

Fig. (4.8). Relation between the minimum net head and the power of turbines [6].

The most common and preferred type of turbine is the Francis turbine. It is used to generate about 60% of the global hydropower in the world making them the most widely used type of turbine. The following charts shows a comparison between all the turbine and how to select each one according to the head and flow rate also the power consumed by the turbine.

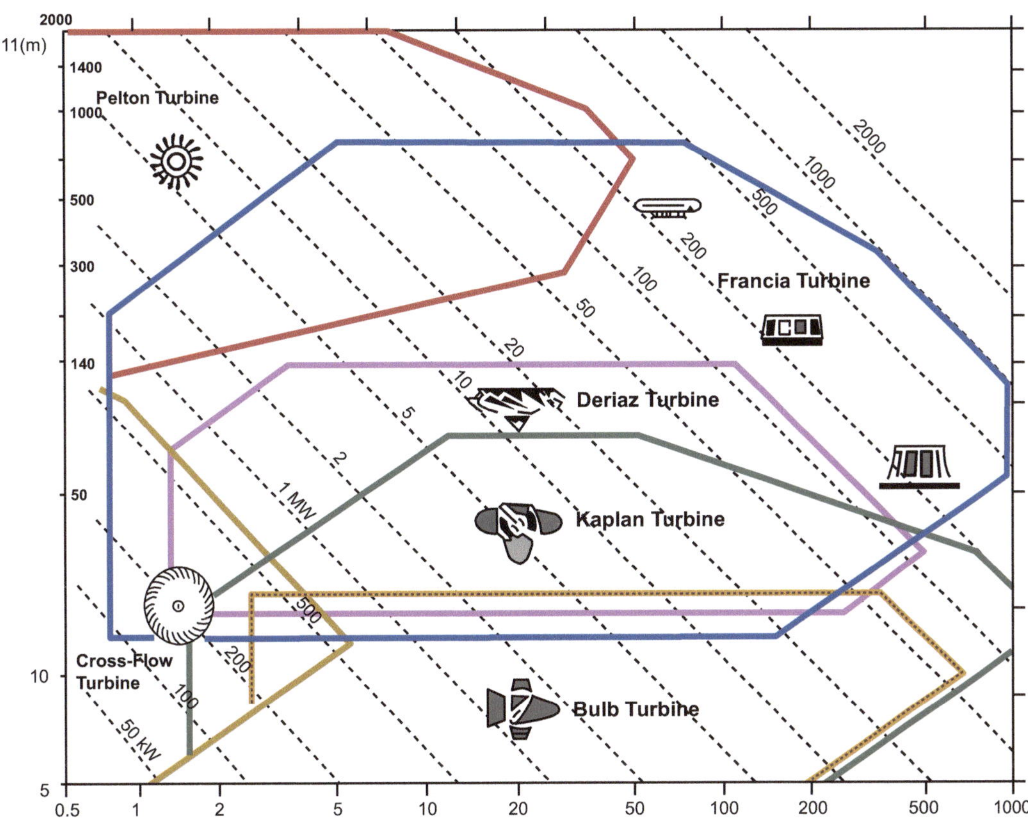

Chart-1-

Turbine selection chart of turbine types. Figure from Heinzmann Hydro Tech Private Limited India. Hydraulic Turbine Classification [5].

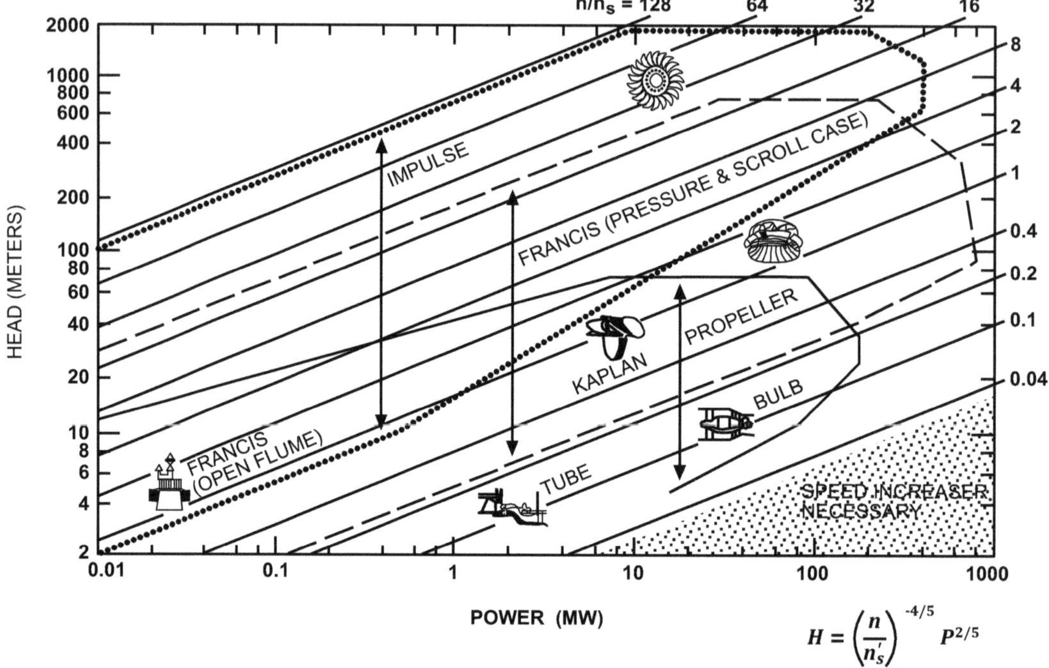

Chart-2-

Specific speed expressed as $nP1/2H{-}5/4$. Source: John S. Gulliver, Roger E.A. Arndt, Hydroelectric Power Stations, In: Encyclopedia of Physical Science and Technology (Third Edition), Academic Press, New York, 2003, Pages 489-504, ISBN 9780122274107, 10.1016/B0-12-227410-5/00321-5. (http://www.sciencedirect.com/science/article/pii/B0122274105003215). Energy Systems - Hydraulic turbines and hydroelectric power plants [5].

CHAPTER 5

Centrifugal and Positive Displacement Pumps

Abstract: A pump is a machine that provides energy to a fluid in a hydraulic system. It assists to increase the pressure energy or kinetic energy, or both, in the fluid by converting the mechanical energy. The basic difference between a turbine and the pump, from a hydrodynamic point of view, is that in the former flow takes place from the high-pressure side to the low-pressure side, whereas in pump flow takes place from the low pressure forwards the higher pressure. Thus in a turbine, there is accelerated flow while in a pump the flow is decelerated. Accelerated flow throughout the hydraulic turbines is less subjected to turbulence therefore the runner passages are relatively short and high efficiency is available for this machine due to reduced values for the friction losses. Decelerated flow throughout the centrifugal pumps is sensitive to separation and vortices therefore impeller passages are relatively long and gradually increased in cross-section area for lowering the friction losses – "centrifugal pumps" efficiency is normally lower comparing to the turbines.

At the beginning of this chapter, one presents a classification of centrifugal pumps, reciprocating pumps – (Fig. **5.1**) and pump turbines. In addition, basic centrifugal pump theory and a brief analysis of the net positive suction head (NPSH) that are very useful for the design and selection of the pumps are detailed. In the next sections similarity laws, specific speed, cavitation and selection of the pumps are available. All these items are illustrated by solved problems.

Chapters on "similarity law, specific speed and cavitation and pumps section" acquiring great efficiency in using the tool of mathematics and at the solved problems are available.

Keywords: Cavitation in pump, Centrifugal pumps, Efficiencies, Force and Power, Head of the pump, Negative suction lift, NPSH required, NPSH, Positive Displacement Pumps, Positive suction lift, Pump Turbine, Reciprocating pump.

5.1. CENTRIFUGAL PUMPS

All types of pumps that depend on the change of momentum during the flow through an impeller across the blades are called centrifugal pumps (C.P.).

The basic principle of the C.P. is that the blades or impellers rotating inside a closed fitting housing draw the liquid into the pump through a central inlet opening and

Jafar Mehdi Hassan, Salman Hussien Omran, Laith Jaafer Habeeb,
Alamaslamani Ammar Fadhil Shnawa & Adrian Ciocănea

by means of centrifugal force or change in momentum in the liquid outward through a discharge outlet at the periphery of the housing. That means changing the kinetic head to pressure head. The general classification of pumps regarding the principle, kind of action upon liquid, motion of working members, and form of working members are shown in Fig (**5.1**).

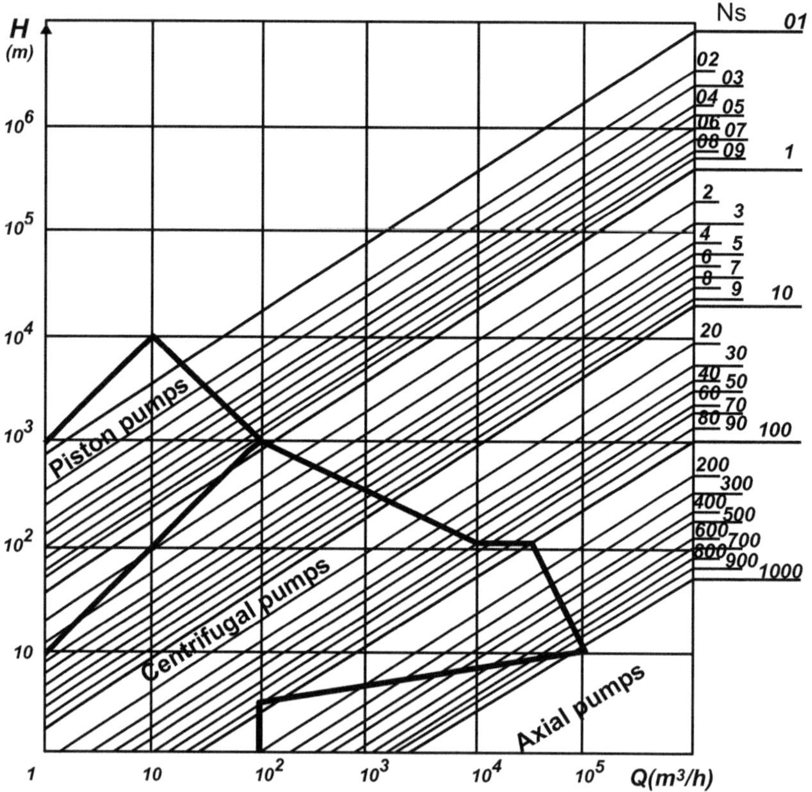

Fig. (5.1). Classification of pumps.

5.2. CLASSIFICATION AND STRUCTURE OF THE CENTRIFUGAL PUMPS

The basic classification of C.P. can consider the working head or hydraulic power of the machine.

1. Working Head - it is the head at which the liquid is delivered by the pump depending on the number of stages (Fig. **5.2** and Table **5.1**):

a) **Low lift centrifugal pumps**: are means to work against heads up to 15 m. impeller is surrounded by a volute and there are no guide vanes. The shaft is generally horizontal and water may enter the impeller from one or both sides depending upon the quantity of water to be delivered.
b) **Medium lift centrifugal pumps**: are used to build up heads as high as 40 m. they are generally provided with guide vanes. Water may enter from one or both sides depending on the quantity to be pumped.
c) **High lift centrifugal pumps**: are employed to deliver liquids at heads above 40 m. high lift pumps are generally multistage pumps because a single impeller cannot easily build up such high pressure. They may be horizontal or verticals the latter being used in deep wells.

Table 5.1. Type of pumps land and working head [4].

Types of Pumps	Working Head m	
Low lift C.P. Medium lift C.P.	Up to 15 15 → 40	} Single-stage
High lift C.P.	>40	Multistage

2. Hydraulic Power - is the absorbed power and represents the energy imparted on the fluid to increase its pressure and velocity, (Table **5.2**).

Table 5.2. Type of pump power and working head.

Types of Pumps	Working Head (m)
Very low C.P Low power C.P. Medium power C.P.	up to 10 1 → 10 10 → 100
High power C.P. Very high power C.P	100 → 1000 >1000 m

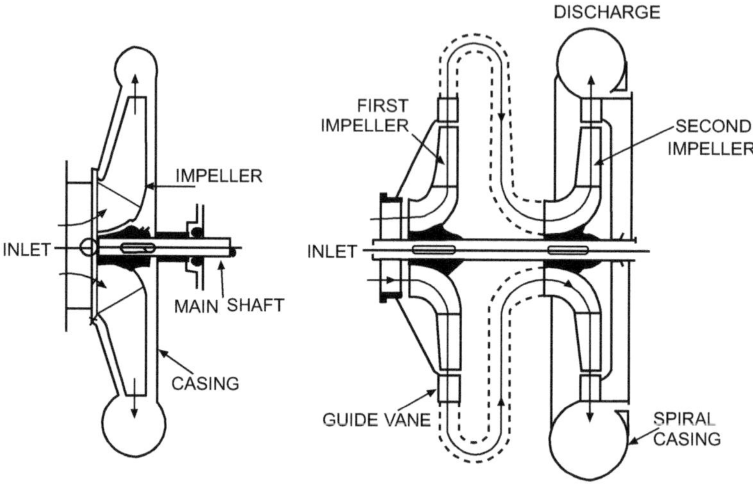

a) Single stage centrifugal pump b) Multi stage (two stage centrifugal pump)

Fig. (5.2). Single and multistage centrifugal pumps [4].

3. The Relative Direction of Flow Through Impeller - According to the flow direction through the impeller, pumps can be considered as: a) Radial flow pump; b) Mixed flow pump; c) Axial flow pump (Fig. **5.3**).

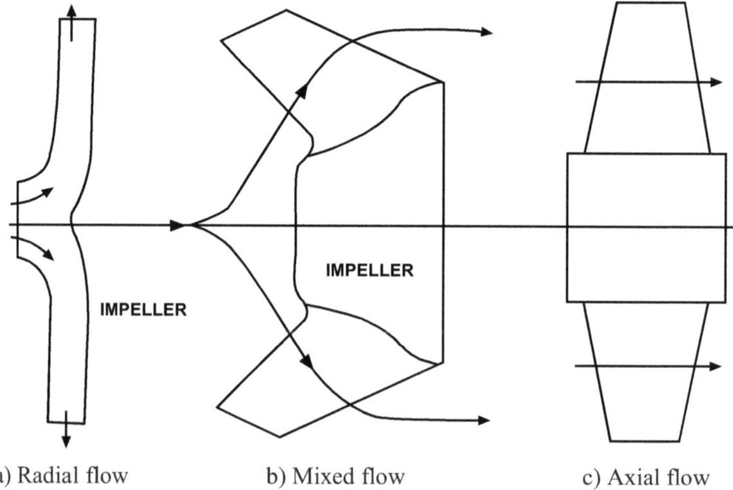

a) Radial flow b) Mixed flow c) Axial flow

Fig. (5.3). Single and multistage centrifugal pumps [4].

4. Impeller Construction - Depending on the properties of liquid to be pumped the centrifugal pumps may have a closed, semi-opened or opened impeller – (Fig. **5.4**). Each type can be manufactured of various materials with or without surface coating to resist chemical or erosion attack of the liquid being pumped:

a) **Closed Impeller Pump**: An ordinary centrifugal pump is equipped with a closed impeller in which the vanes are covered with shrouds on both sides (Fig. **5.4a**). This type is meant to handle non – viscous liquid such as ordinary water, hot water, hot oil and chemicals like acids, *etc*. material of the impeller should be selected according to the chemical cast steel impeller is recommended. For example, hot water temperature exceeding 150°C, cast steed impeller is recommended.

b) **Semi-Open Impeller Pump**: The impeller is provided with a shroud on one side only (Fig. **5.4b**). This pump is used for viscous liquids such as sewage water, paper pulp, sugar molasses, *etc*. In order to minimize the chances of impeller getting clogged, the number of vanes is reduced and their height is increased.

c) **Open Impeller Pump**: The impeller is not provided with any shroud (Fig. **5.4c**). Such pumps are used in dredgers and elsewhere for handling mixture of water, sand, pebbles and clay, in which the solid contents may be as height as one part in four. The impeller has very rough duty to perform. It is generally mads of forged steel its life depends upon the material handled may be 40 or 50 hours or in some case from 500 to 1000 houres.

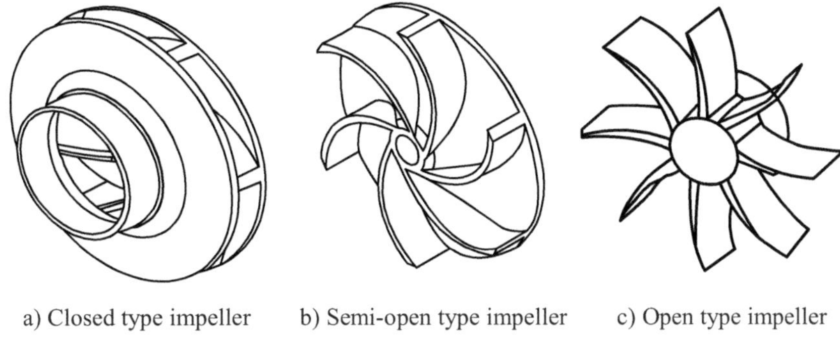

Fig. (5.4). Types of impellers [8].

5. Number of Entrance to the Impeller - The pump may be single or double entry type. In single entry, the liquid inters the impeller from suction pipe on one side, while for the double entry both sides are available.

6. Disposition of the Shaft and Number of Impeller per Shaft - The shaft of the pump may be disposed horizontally or vertically according to the installation conditions. The number of stages of the C.P. is related to the working head requested by the pumping application:

Single Stage C.P. - it has one impeller keyed to the shaft.

Multi Stage C.P. - it has two or more impellers keyed to single shaft enclosed in the same case.

7. Specific Speed - N_s - Is defined as the speed of a geometrically similar pump when delivering 1 m^3/s against a head of 1 meter:

$$N_s = \frac{NQ^{1/2}}{H^{3/4}} \quad \textbf{(5.1)}$$

where: N – rotating speed (rpm); H - head of the pump (m); Q – flow rate (m^3/s).

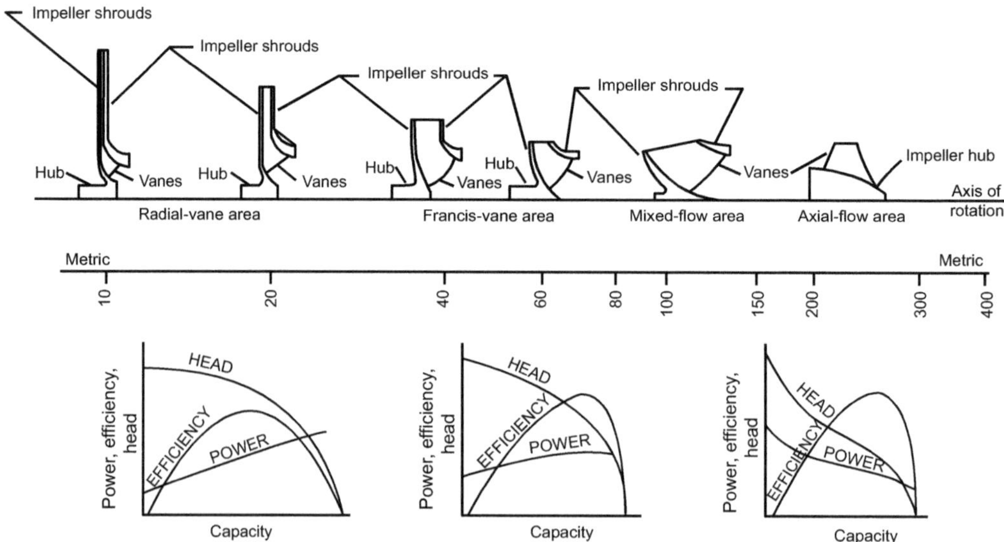

Fig. (5.5). General impeller types according to specific speed Ns and characteristic curves [9].

Impeller types according to specific speed N_s and characteristic curves are presented in Fig. (**5.5**).

Note: The value of H used in this equation considers a single stage for the pump. For multistage pumps H should be divided by the number of stages. Also, flow rate used considers a single suction. For double suction should be divided by 2.

8. Casing of Centrifugal Pumps - The role of the pump casing is to convert the kinetic energy to pressure energy. Pump casing should be so designed as to minimize the loss of kinetic head transferred by the impeller to the liquid Efficiency of the pump depends on the type of casing (Fig. **5.6**).

a) Volute Casing (Spiral Casing):-
In a spiral casing the gradual increase in the area of flow decrease the velocity of the liquid and accordingly the pressure value is increasing. For this casing type – (Fig. **5.6a**) - ca considerable loss takes place due to the formation of eddies;

b) Vortex Casing:-
It is an improved type of a volute casing in which the spiral casing is combined with circular chamber as shown in Fig. (**5.6b**). In this configuration the eddies are less intense and the efficiency is increased;

c) Volute Casing with Guide Blades:-
For this casing the impeller is surrounded by guide blades – (**Fig. 5.6c**) which are arranged at an optimum angle, depending of the absolute angle of the liquid velocity at the outlet of the impeller. The liquid enters without shock in the passages of increasing area and reaches the delivery pipe. The ring of the guide blades is called diffuser and is very efficient.

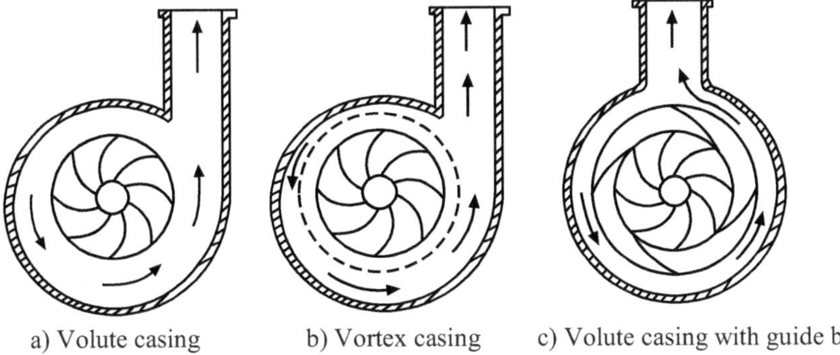

Fig. (5.6). Types of casing [4].

5.3. THEORY OF CENTRIFUGAL PUMPS

Centrifugal pumps are dynamic pumps. The basic theory of a centrifugal pump can be developed by considering the average one - dimensional fluid flow between the inlet and the outlet sections of the impeller as the blades rotate.

Let the points on the liquid path through the pump from suction inlet, impeller inlet to the impeller outlet, be denoted by i, 1, 2 and d, as shown in Fig. (**5.7**) below:

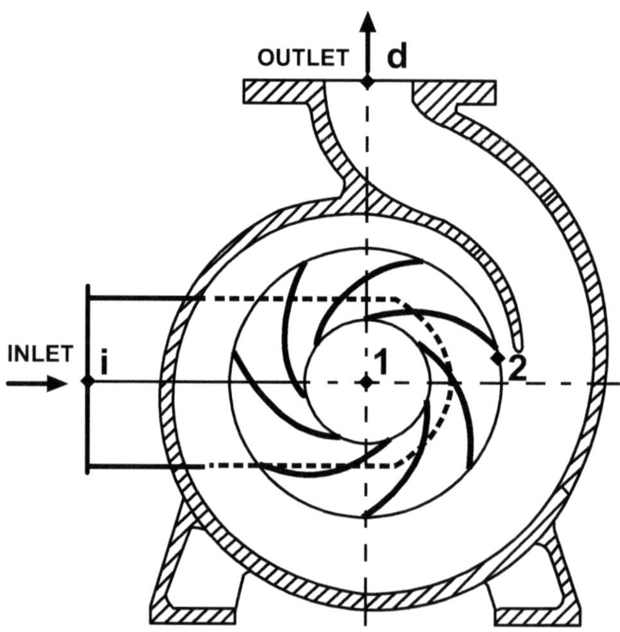

Fig. (5.7). Liquid flow through a centrifugal pump [8].

a) Between i & 1, through the stationary suction pipe V_i, V_l the absolute velocities.

$$\frac{V_i^2}{2g} + \frac{P_i}{\gamma} + Z_i = \frac{V_1^2}{2g} + \frac{P_1}{\gamma} + Z_1 + H_{L_{i-1}} \quad \textbf{(5.2)}$$

b) Between 1 & 2 through the movable impeller and W_1,W_2 the relative velocities.

$$\frac{W_1^2}{2g} - \frac{U_1^2}{2g} + \frac{P_1}{\gamma} + Z_1 = \frac{W_2^2}{2g} - \frac{U_2^2}{2g} + \frac{P_2}{\gamma} + Z_2 + H_{L_{1-2}} \quad \textbf{(5.3)}$$

c) Between 2 & d through the casing to the discharge section and V_2, V_d absolute velocities

$$\frac{V_2^2}{2g}+\frac{P_2}{\gamma}+Z_2=\frac{V_d^2}{2g}+\frac{P_d}{\gamma}+Z_d+H_{L_{2-d}} \quad \textbf{(5.4)}$$

From 5.2-5.4 we get:

$$\frac{V_2^2-V_3^2}{2g}+\frac{W_1^2-W_2^2}{2g}+\frac{U_2^2-U_1^2}{2g}=\left[\left(\frac{V_d^2}{2g}+\frac{P_d}{\gamma}+Z_d\right)-\left(\frac{V_i^2}{2g}+\frac{P_i}{\gamma}+Z_i\right)\right]+\left(H_{L_{i-1}}+H_{L_{1-2}}+H_{L_{2-d}}\right) \quad \textbf{(5.5)}$$

The first terms on the right hand side by definition are: the total manometric head - equal to H_{mano}- and the total losses due to fluid inside the pump only.

$$\frac{V_2^2-V_1^2}{2g}+\frac{W_1^2-W_2^2}{2g}+\frac{U_2^2-U_1^2}{2g}=H_{mano}+\Delta H_{mano} \quad \textbf{(5.6)}$$

This relation it is known as the fundamental equation of C.P. Considering the losses of head in the pump, its efficiency:

$$\eta_{mano}=\frac{H_{mano}}{H_{mano}+\Delta H_{mano}}=manometric\ efficiency$$

$$or \quad H_{mano}+\Delta H_{mano}=\frac{H_{mano}}{\eta_{mano}}$$

Therefore

$$\frac{V_2^2-V_1^2}{2g}+\frac{W_1^2-W_2^2}{2g}+\frac{U_2^2-U_1^2}{2g}=\frac{H_{mano}}{\eta_{mano}} \quad \textbf{(5.7)}$$

This equation can be simplified by substituting for W_1 and W_2 from velocity triangle at inlet and outlet of the impeller – (Fig. **5.8**).

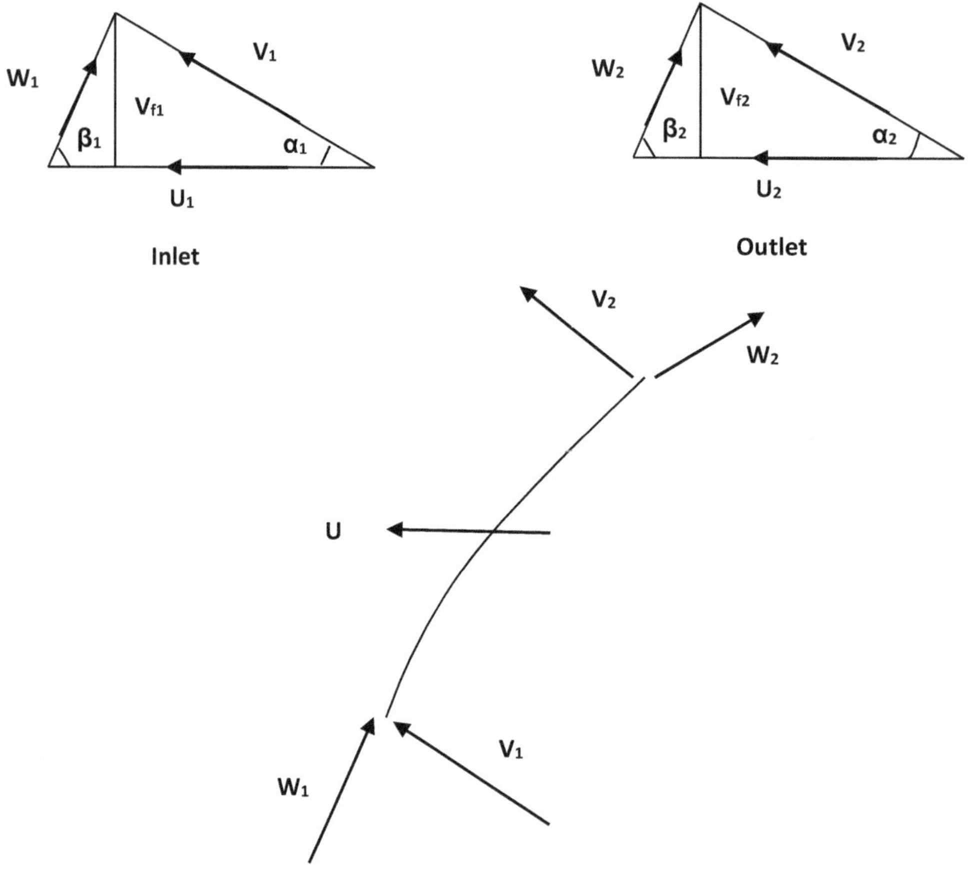

Fig. (5.8). Velocity triangles at input and output of the impeller.

$$W_1{}^2 = U_1{}^2 + V_1{}^2 - 2U_1V_1\cos\alpha_1$$

$$W_2{}^2 = U_2{}^2 + V_2{}^2 - 2U_2V_2\cos\alpha_2$$

$$W_1{}^2 - W_2{}^2 = U_1{}^2 - U_2{}^2 + V_1{}^2 - V_2{}^2 - 2U_1V_1\cos\alpha_1$$

$$+2U_2V_2\cos\alpha_2 \quad \textbf{(5.8)}$$

From equations (5.7 and 5.8) we get

$$\frac{H_{mano}}{\eta_{mano}} = \frac{2U_2V_2\cos\alpha_2 - 2U_1V_1\cos\alpha_1}{2g}$$

Denoting $V\cos\alpha = V_u$ and considering $\alpha_1 = 90°$ as the ideal value for absolute angle of the liquid velocity at the impeller inlet (radial flow at inlet) $\cos\alpha_1 = 0$ manometric efficiency is:

$$\therefore \eta_{mono} = \frac{gH_{mano}}{U_2Vu_2} \quad \textbf{(5.9)}$$

5.4. HEAD OF THE CENTRIFUGAL PUMPS

The term head of a pump stands for the following (Fig. **5.9**):

a) Static Head: - it is difference between elevation of upper and lower reservoirs, or the sum of the suction and delivery head:

$$H_{static} = H_{\text{suction}} + H_{\text{delivery}} \quad \textbf{(5.10)}$$

b) Manometric Head: - it is the head measured across the pump inlet and outlet flanges. It expresses the increase in pressure energy per unit weight of liquid handled by the impeller:

$$\eta_{mano} = \frac{P_d - P_s}{\gamma} + h_g$$
$$= H_{mano(d)} - H_{mano(s)} + h_g \quad \textbf{(5.11)}$$

Where h_g - The vertical distance between the pressure tapping for the suction and delivery gauges.

Note: the suction pressure is considered of negative value (-) if the level of free surface of the liquid is below than the impeller axis and (+) when in the opposite case.

The manometric head includes static head and all the losses which the pump has to compensate except the kinetic head.

c) Total or Effective Head: - it is the actual head against which the pump has to work:

$$H_p = \frac{P_d - P_s}{\gamma} + h_g + \frac{V_d^{\,2} - V_s^{\,2}}{2g} \qquad \textbf{(5.12)}$$

Where V_d - Velocity of flow at delivery pipe; V_s - Velocity of flow at suction pipe.

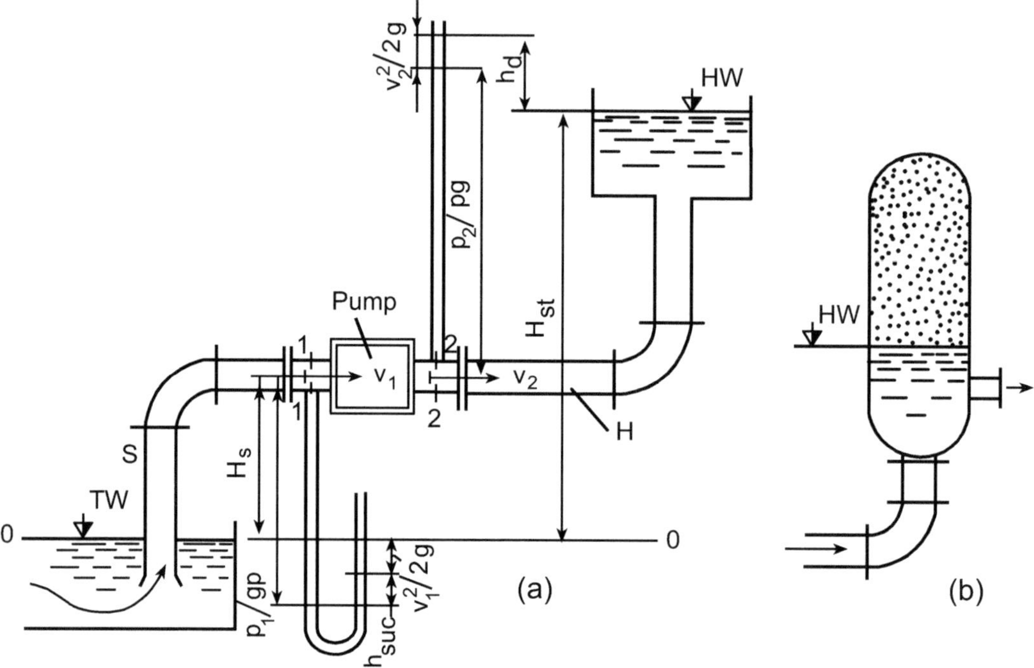

Fig. **(5.9).** Schematic diagram of pump installation [6].

The difference between the manometric head and the total head is the kinetic head; in practical we consider that manometric head is equal to the total head:

$$\therefore\ H_{mano} = H_{total} = H_{static} + H_L$$

Where H_L - The total losses of head

5.5. FORCE AND POWER OF CENTRIFUGAL PUMPS

The force due to momentum change in the impeller is:

$$F = \dot{m}(V_{u2} - V_{u1}) \qquad \dot{m} = \rho Q$$

$$\therefore Power = F.U \ power \text{ delivered by the impeller}$$

The output power of the pump $= \gamma Q H_p$

$$\therefore \eta_{mano} = \frac{\rho Q (V_{u2} U_2 - V_{u1} U_1)}{\gamma Q H}.$$

For radial inlet flow $\alpha_1 = 90°$ & $\cos \alpha_1 = 0$ (no loss condition):

$$\therefore \eta_{mano} = \frac{V_{u2} U_2}{g H} \qquad \textbf{(5.13)}$$

$$\frac{V_{u2} U_2}{g} = H_{mano}$$

is the work done per unit mass of liquid.

5.6. EFFICIENCIES OF CENTRIFUGAL PUMPS

For C.P the overall efficiency takes into account the hydraulic losses, mechanical losses and the internal leakage of the pump, mainly due to dynamic sealing of the impeller.

a) Overall Efficiency:

$$\eta_{overall} = \frac{Fluid\ power\ outlet}{Power\ input\ by\ the\ shaft}$$

$$\eta_{o.} = \frac{\gamma Q H_{mano}}{SKW} \qquad \textbf{(5.14)}$$

$$P_{shaf} = P_{input\ to\ the\ impeller} + P_{leakage} + P_{mech.loss}$$

$$P_{input\ to\ the\ impeller} = energy\ given\ to\ impeller\ per\ kW$$

$$= \frac{V_{u2}U_2}{g}$$

b) Mechanical Efficiency: it is the ratio of power delivered by the impeller to the fluid to the power input to the shaft:

$$\eta_{mech} = \frac{SKW - P_{mech.Loss}}{SKW} \tag{5.15}$$

$P_{mech.loss}$ – Power required to overcome all mechanical losses *i.e.* friction losses, bearing and glands losses.

c) Volumetric Efficiency:

$$\eta_Q = \frac{Q}{Q+\Delta Q} \tag{5.16}$$

Where: Q – discharge developed by the pump; ΔQ – amount of leakage.

d) Manometric Efficiency:

$$\eta_{mano} = \frac{actual\ measured\ head\ or\ gross\ lift}{head\ impeller\ to\ the\ fluid\ by\ impeller}$$

$$\eta_{mano} = \frac{H_{static} + \sum H_L}{\frac{V_{u2}U_2}{g}} \quad or \quad = \frac{gH}{V_{u2}U_2} \tag{5.17}$$

This also known as hydraulic efficiency:

$$\eta_o = \eta_{mech} \cdot \eta_Q \cdot \eta_{mano} \tag{5.18}$$

5.7. FLOW RATE THROUGHOUT THE CENTRIFUGAL PUMP

According to the geometry of the impeller – (Fig. **5.10**) and considering incompressible fluid passing through the pump, the flow rate Q is the quantity of flow (m^3/s) flowing through the impeller (neglecting the leakage):

$$Q = \pi D_1 B_1 V_{f1} = \pi D_2 B_2 V_{f2} \tag{5.19}$$

Where D & B the diameter and the width or breadth of the impeller.

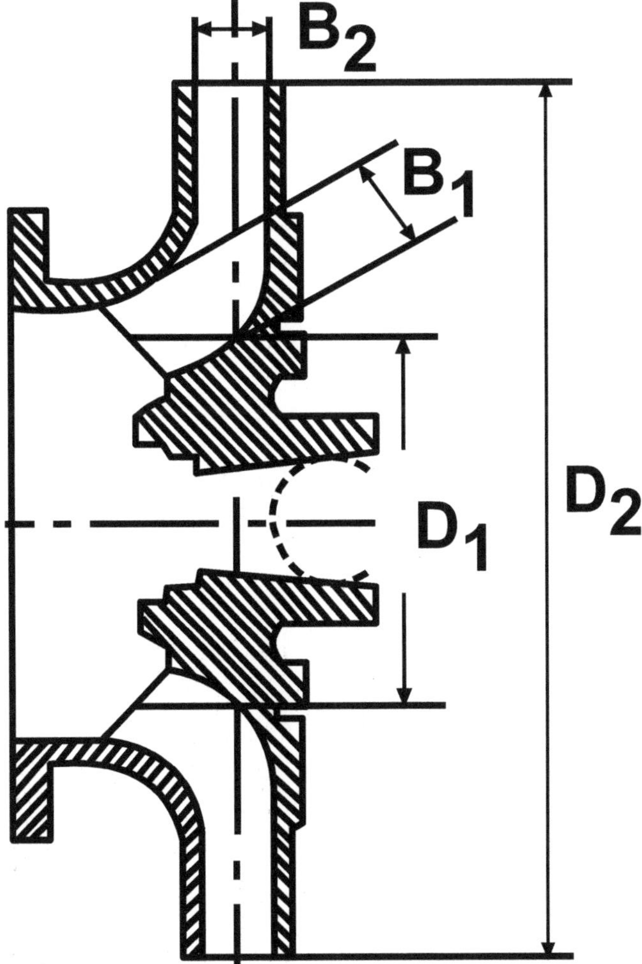

Fig. (5.10). Geometry of the impeller [8].

For constant radial flow velocity V_f :

$$D_1 B_1 = D_2 B_2$$

5.8. NET POSITIVE SUCTION HEAD (NPSH)

It is a term describing the conditions related to cavitation inception, which is undesired and harmful for the operation of centrifugal pumps. In analyzing a pump operating in a system to determine if cavitation is likely to occur, there are two aspects of NPSH to consider, $NPSH_a$ and $NPSH_r$.

5.8.1. Net Positive Suction Head Available ($NPSH_a$)

$NPSH_a$ is the absolute pressure available at the suction section of the pump over the vaporization pressure of the fluid. $NPSH_a$ is a function of the suction system and is regardless of the pump type working in the system. It is computed as follows:

$$\mathrm{NPSH_a} = \mathrm{H_a} \mp \mathrm{H_s} - \mathrm{H_f} - \mathrm{H_v} \qquad \textbf{(5.20)}$$

Where

$\mathrm{H_a}$ − Atmospheric pressure (m);
$\mathrm{H_s}$ − Suction head (m) + reservoir above the pump;
- reservoir below the pump;
$\mathrm{H_f}$ − Losses due to friction in suction line (m);
$\mathrm{H_v}$ − Vapor pressure of liquid as a function of temperature (m).

Computation of Net Positive Suction Head Available ($NPSH_a$)

$NPSH_a$ at pump suction depends on fluid's nature and pressure, temperature, vapor tension, altitude, diameter and shape of the pipes, *etc.* – (Fig. **5.11**). In order to have the installation running properly, it is mandatory to have the NPSH available the pump's suction higher than the NPSH necessary for the same pump:

NPSH available > NPSH required.

The security range is to be between 0.5 to 1 m head depending on the pump.

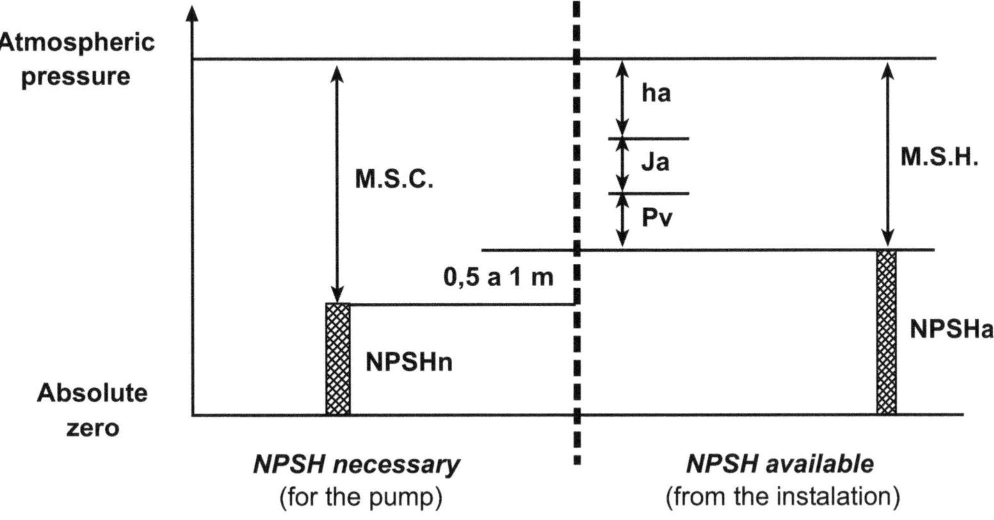

Fig. (5.11). Relation between $NPSH_a$, $NPSH_r$, MSC and MSH [10].

where: h_a – suction geometric head; J_a – suction piping network's total friction losses [11].

$$\text{M.S.C.} = 10\left(\frac{\text{Pa}-\text{Pv}}{\gamma}\right) + \frac{V_a{}^2}{2g} \tag{5.21}$$

where:

M.S.C. – Maximum Suction Capacity (m) – it is given in meters of water and indicates the height above which a pump is able to draw water and pump normally; $NPSH_a$, $NPSH_r$ (m) – Net Positive Suction Head available/required (necessary); P_a – absolute pressure at the pump's suction (N/m^2); P_V- vapor tension of the pumped fluid (N/m^2); γ – Specific Gravity of the liquid (pure water γ is $1 g/cm^3$); V_a – liquid velocity at the pump suction (m/s); g – gravity acceleration (m/s^2);

5.8.2. Net Positive Suction Head Required ($NPSH_r$)

$NPSH_r$ is the minimum pressure required at the suction section of the pump to keep the pump from cavitation. $NPSH_r$ is depending only on the pump inlet design, and is independent of the suction piping system. The value for $NPSH_r$ is provided by the pump manufacturer - it depends of the pump type, the impeller diameter, the flow and the pump speed. The $NPSH_r$ (requested) of a pump, given in meters of liquid, indicates the minimum absolute pressure necessary at the pump's suction for correct running. $NPSH_r$ enables the pump's MSC (Maximum Suction Capacity) to be calculated. The MSC is given in meters of water and indicates the height above which a pump is able to draw water and pump normally [11] – (Fig. **5.12**).

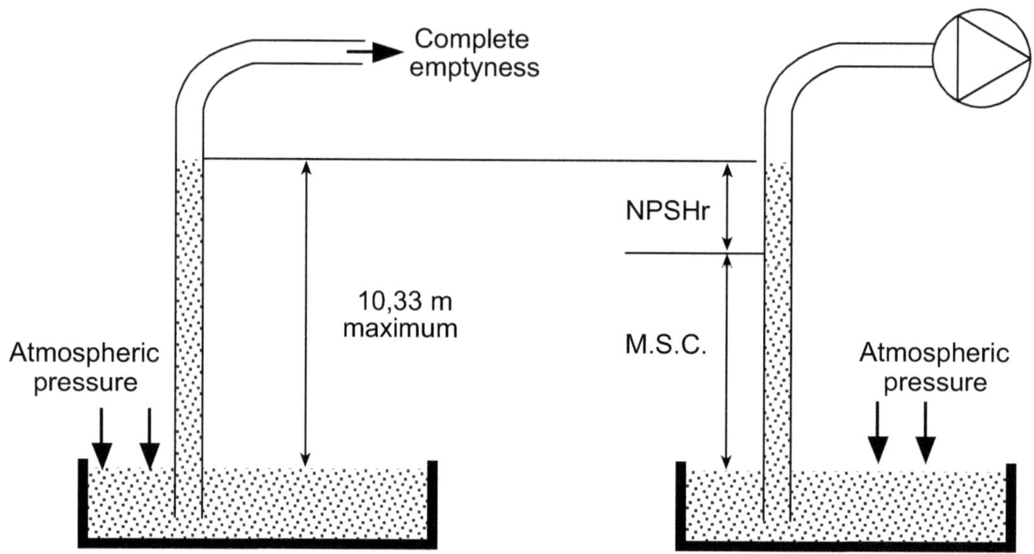

Fig. (5.12). Relation between NPSHr and MSC [10].

According to the installation structure *i.e.* the pressure value in the suction tank, there are two relations available:

- Pump drawing cold water from a tank at atmospheric pressure:

M.S.C. = Patm - NPSHr

- Pump sucking from a pressurized tank

$$\mathrm{M.S.C.} = 10\left(\frac{\mathrm{Po}-\mathrm{Pv}}{\gamma}\right) \quad \textbf{(5.22)}$$

M.S.C. - Maximum Suction Capacity;
P_{atm} - absolute pressure in the suction tank (in bar);
P_v - Vapor tension of the pumped liquid at the pumped temperature (in bar);
γ - Specific Gravity - pure water γ is 1 g/cm³;

5.8.3. Negative Suction Lift

According to the position of the pump in the installation, negative suction lift considers the tank below the pump – (Fig. **5.13**).

NPSHa: Net **P**ositive **S**uction **H**ead **a**vailable, depends on the installation.
NPSHr: Net **P**ositive **S**uction **H**ead **r**equested, depends on the pump (given by the pump manufacturer).
MSC: Maximum **S**uction **C**apacity.

TH: Total **H**ead.
=> **NPSHa** $= (P_{atm} - P_V) / \gamma - H_s - J_s$
=> **TH** $= H_s + J_s + H_d + J_d + P_r$
J_s : friction loss in the suction pipe (m)
J_d : friction loss in the delivery pipe (m)
=> **MSC** $= P_{atm} - NPSH_r$

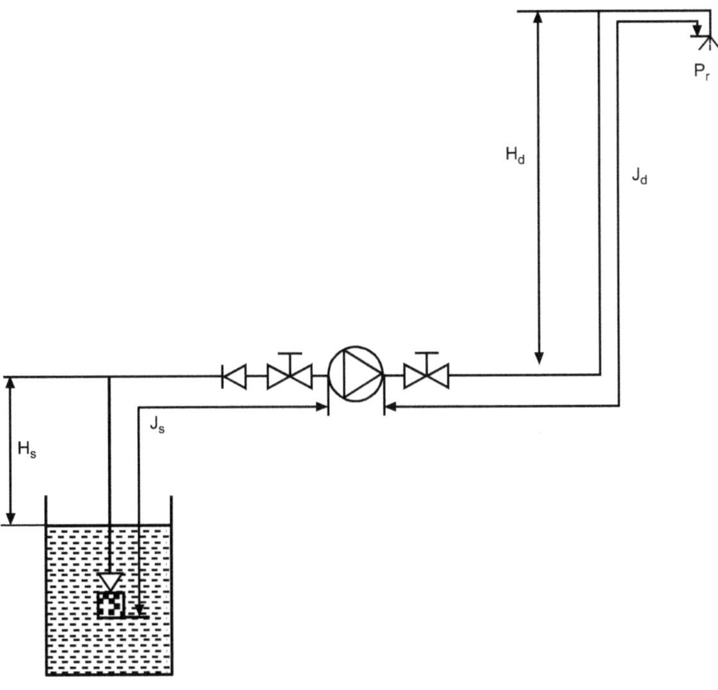

Fig. (5.13). Negative suction lift [10].

CAUTION:

NPSHa > NPSHr – minimum 0.5 – 1 m
CAVITATION risk if **NPSHa**< NPSHr
MSC > H_a+J_a – minimum 0.5 to 1m
If temperature increases, **MSC** decreases
If altitude increases (Patm decreases), **MS** decreases

5.8.4. Positive Suction Lift

According to the position of the pump in the installation, positive suction lift considers the tank on top of the pump – (Fig. **5.14**)

NPSHa: Net **P**ositive **S**uction **H**ead **a**vailable, depends on the installation.
NPSHr: Net **P**ositive **S**uction **H**ead **r**equested, depends on the pump - given by the pump manufacturer.
MSC: Maximum **S**uction **C**apacity

TH: Total **H**ead
=> $\mathbf{NPSH_a} = (P_{atm} - P_v) / \gamma - H_s - J_s$;
=> $\mathbf{TH} = H_s + Js + H_d + J_d + P_r$
=> $\mathbf{MSC} = P_{atm} - NPSH_r$

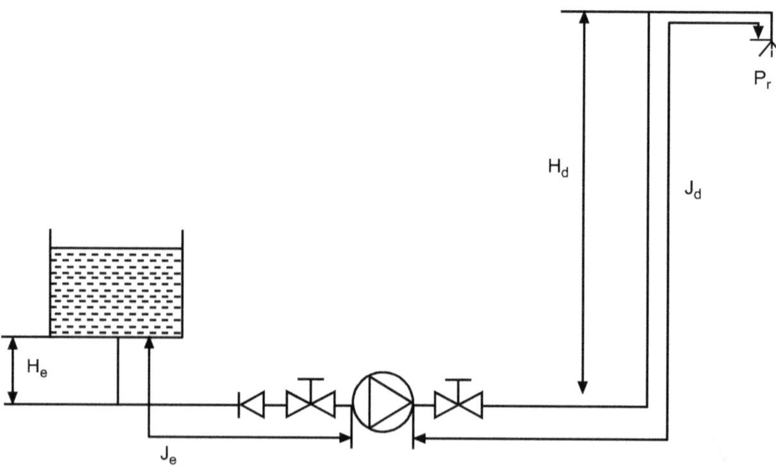

Fig. (5.14). Positive suction lift [10].

CAUTION:

$NPSH_a > NPSH_r$ - minimum 0,5 to 1 meter

CAVITATION risk if $NPSH_a < NPSH_r$

MSC > $H_a + J_a$ - minimum 0,5 to 1 meter

If temperature increases, **MSC** decreases

If altitude increases (P_{atm} decreases), **MSC** decreases

5.9. CAVITATION IN PUMP

When the liquid is flowing in the pump, it is possible that the pressure at any part of the pump may fall below the vapor pressure. In this case the liquid will vaporize and the flow will no longer continuous. The vaporization will appear in the form of bubbles. When passing through a region of high pressure these bubbles collapses on the metallic surface - such as tips of impellers blades - and cavities are formed. This phenomenon is known as cavitation. To prevent the cavitation phenomenon the basic relation that should be valid for the pumping process is:

$NPSH_a > NPSH_r$

The risk of cavitation in pumping systems can be reduced by:

- lowering the pump position compared to the water surface from the open system;
- increasing the system pressure of the closed system;
- shortening the suction line to reduce the friction loss;
- increasing the suction lines cross-section area to reduce the fluid velocity and therefore minimization of friction;
- avoiding pressure drops coming from bends and other obstacles in the suction line;
- lowering fluid temperature to reduce vapor pressure;

The relation between the head, NPSH, and the flow rate is presented in Fig. (**5.15**), where a 3% head drop is considered as cavitation inception.

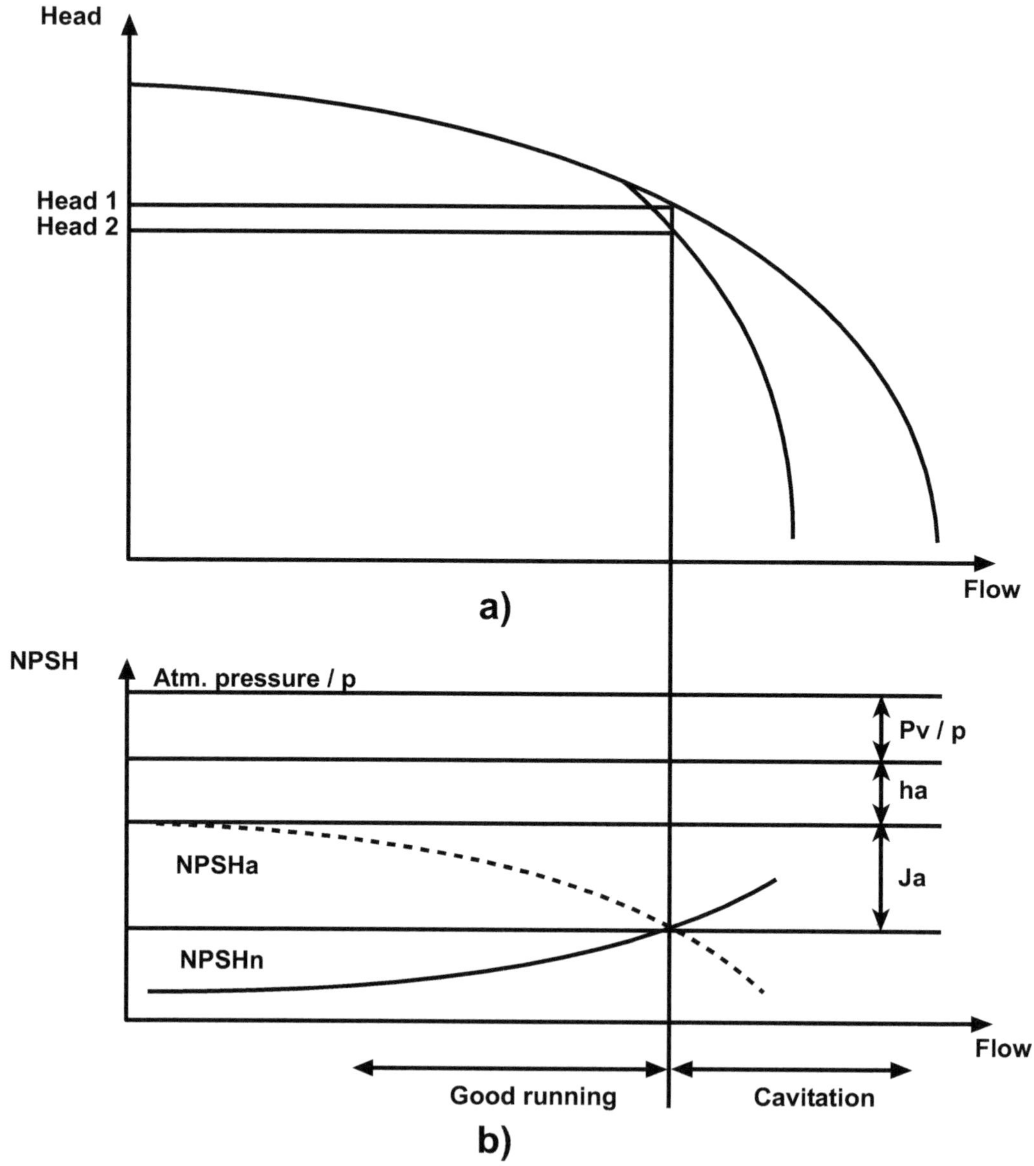

Fig. (5.15). (a) The relation between the head and flow rate. **(b)** The relation between the NPSH and flow rate [10].

5.10. SIMILARITY LAWS

Similarity laws are useful tools to find the characteristic curves and dimensions for turbopumps within the same family. Considering the three dimensionless coefficients for a pumps [8]:

$$\psi = \frac{gH}{U_2^2};\ \varphi = \frac{H}{U_2^2 R_2^2}\ ;\ \pi = \frac{P'}{\rho U_2^3 R_2^2};\ \eta_i = \frac{\varphi\psi}{\pi} \qquad \textbf{(5.23)}$$

as: ψ head coefficient φ flow rate coefficient π dimensionless power coefficient (only for the internal power); η_t = internal efficiency; one obtain relations useful in practice a minor change in head on capacity is needed.

a) Effect of Speed Variation – the case of a single pump (constant impeller diameter) that works with two different rotational speeds N and N', under two similar working conditions:

$$\frac{Q}{Q'} = \frac{N}{N'} \ldots\ldots\ldots (1)$$

$$\frac{H}{H'} = \left(\frac{N}{N'}\right)^2 \ldots\ldots\ldots (2)$$

$$\frac{P}{P'} = \left(\frac{N}{N'}\right)^3 \ldots\ldots\ldots (3)$$

Note. The gap between the two rotational speeds must be considered inside the interval $N \in [0.8\,N';\ 1.2\,N']$.

b) Effect of Diameter Variation – the case of constant rotational speed:

$$\frac{Q}{Q'} = \left(\frac{D}{D'}\right)^2 \ldots\ldots\ldots (1)$$

$$\frac{H}{H'} = \left(\frac{D}{D'}\right)^2 \ldots\ldots\ldots (2)$$

$$\frac{P}{P'} = \left(\frac{D}{D'}\right)^4 \dots\dots\dots (3)$$

c) Pump Modeling – the case of two pumps working in similar hydrodynamic conditions, where specific speed N_s is used:

$$N_s = \frac{N\sqrt{Q}}{H^{\frac{3}{4}}} \dots\dots\dots (1)$$

$$or\ (N_s)_m = (N_s)_p \dots\dots\dots (2)$$

For the same condition the speed, flow rate and the power:

$$\frac{N_m}{N_p} = \frac{D_p}{D_m}\sqrt{\frac{H_m}{H_p}} \dots\dots\dots (3)$$

$$\frac{Q_m}{Q_p} = \left(\frac{D_m}{D_p}\right)^2 \sqrt{\frac{H_m}{H_p}} \dots\dots\dots (4)$$

$$\frac{P_m}{P_p} = \left(\frac{D_m}{D_p}\right)^2 \left(\frac{H_m}{H_p}\right)^{3/4} \dots\dots\dots (5)$$

$P = \gamma QH$

where: p – pump prototype; m – pump model

d) Minimum Value for the Pump Starting Speed - a centrifugal pump will start delivering the liquid only when the head developed by it equals the manometric head at the time of start.

$$\frac{U_2{}^2 - U_1{}^2}{2g} \geq H_{mano} \qquad U = \frac{\pi DN}{60}$$

$$\therefore \frac{U_2{}^2 - U_1{}^2}{2g} = H_{mano} = \frac{V_{u2}U_2}{g}\eta_{mano} \qquad \textbf{(5.24)}$$

e) Minimum Diameter of Impeller: Normally at the starting of design of an impeller the assumption as follows:

$$D_2 = 2D_1$$

$$or \quad \frac{\left(\frac{\pi D_2 N}{60}\right)^2 - \left(\frac{\pi D_1 N}{60}\right)^2}{2g} = H_{mano}$$

$$D_2 = \frac{97.7\sqrt{H_{mano}}}{N} \qquad \textbf{(5.25)}$$

$$D_2 = outlet\ diameter\ (m)$$

$$D_1 = inlet\ diameter\ (m)$$

The following figures describes the characteristics and the working principles of the pump.

5.11. CHARACTERISTIC CURVES FOR VARIOUS WORKING CONDITIONS

Centrifugal, mixed and axial pumps have characteristic curves H(Q), P(Q) and η(Q) with specific shapes according to hydraulic particularities (Fig. **5.16**).

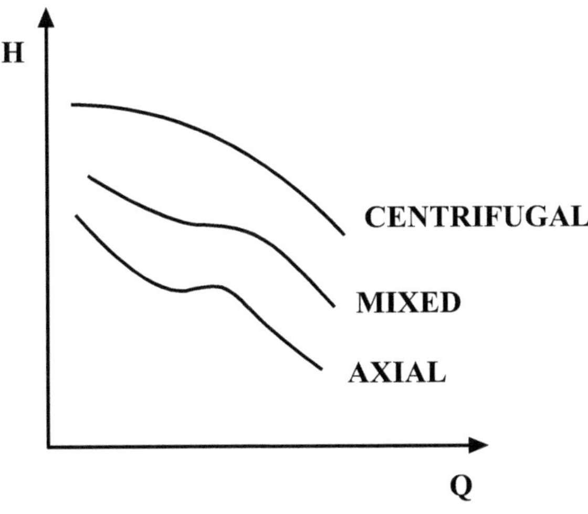

a)

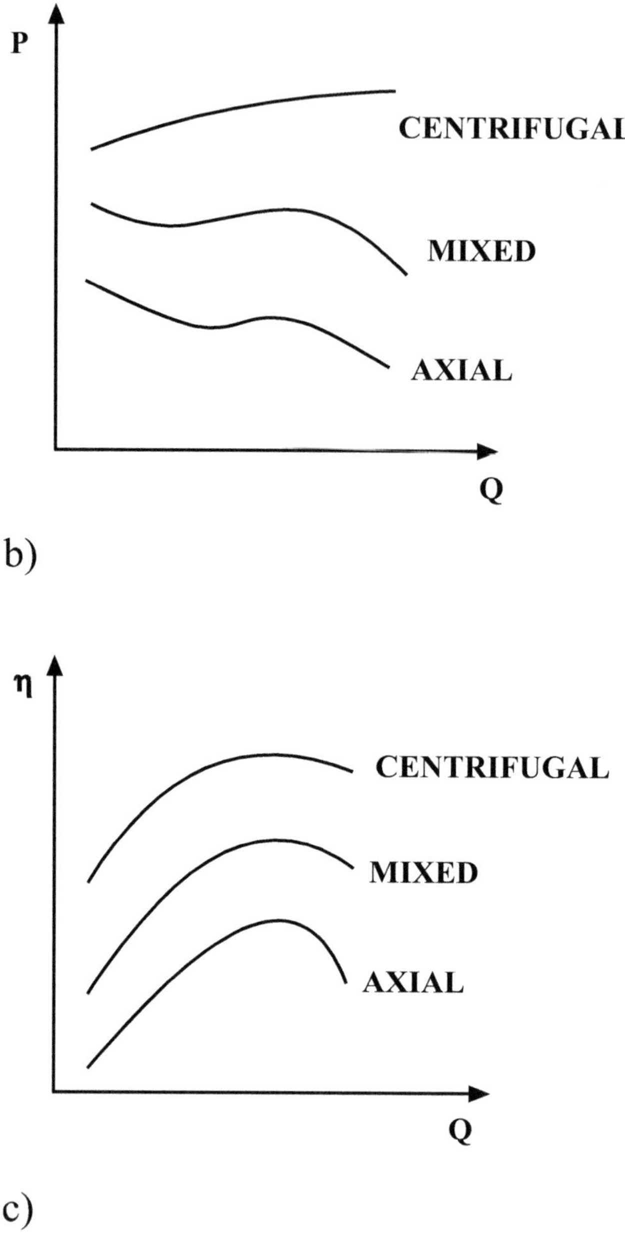

Fig. (5.16). Characteristic curves for axial, mixed and centrifugal pumps. a) H-Q characteristic curves; b) P-Q characteristic curves; c) η-Q characteristic curves [8].

For centrifugal pumps some basic characteristic curves are presented according to the pumping systems.

a) Pump's hydraulic characteristic for closed loop (Fig. **5.17a**) – friction losses curve has no geometrical head.

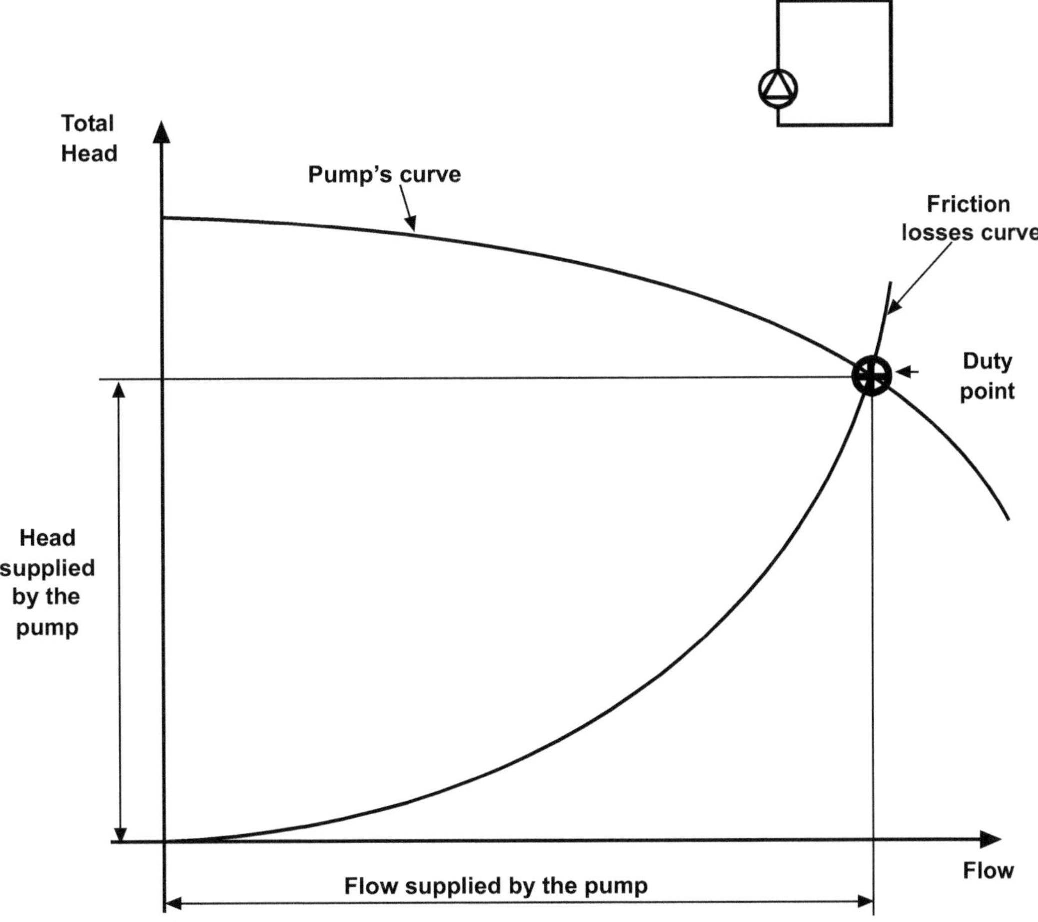

Fig. (5.17a). Pump's hydraulic characteristic - closed loop [10].

b) Pump's hydraulic characteristic for open loop (Fig. **5.17b**) – friction losses curve has geometrical head.

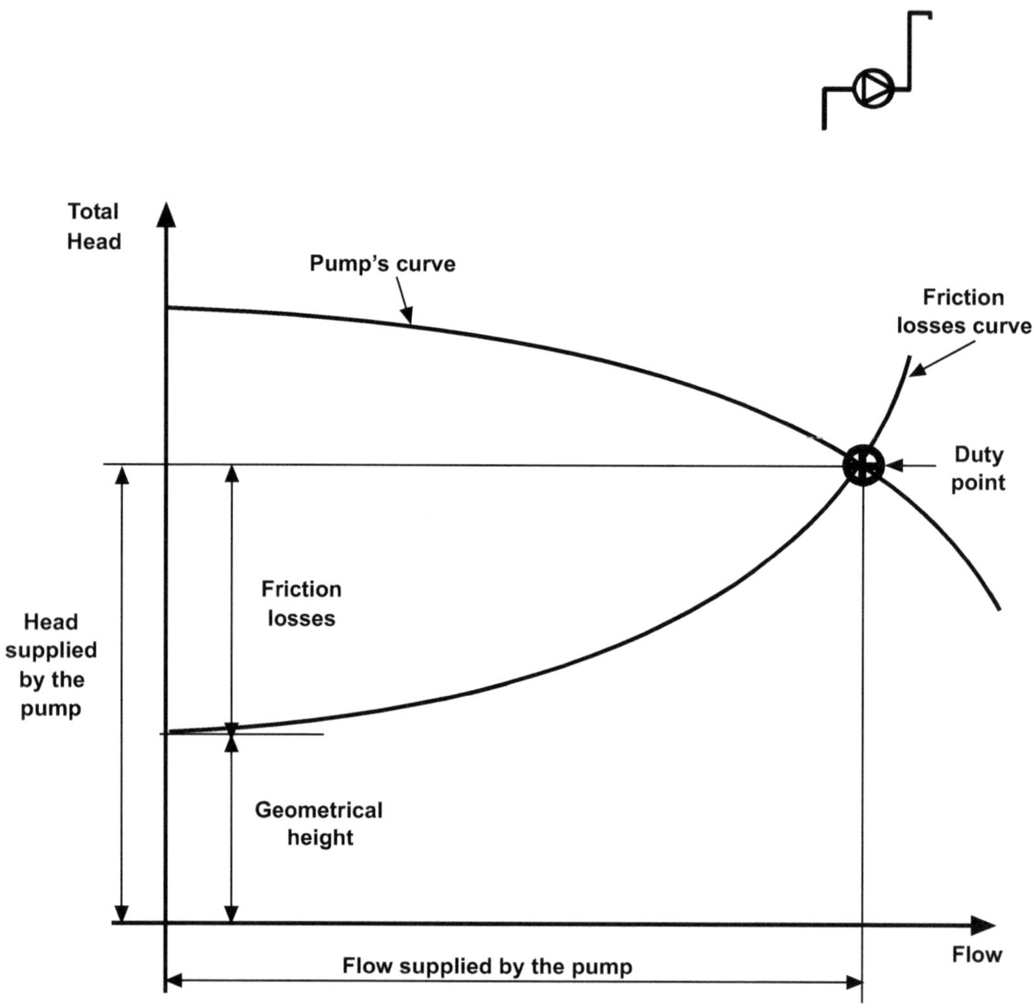

Fig. (5.17b). Pump's hydraulic characteristic - open loop [10].

When a pump is selected to work a pumping system the friction losses are computed in order to minimize the energy consumption. If friction losses are overestimated then over-dimensioning of the pump occur, *e.g.* too large flow, higher power consumption, overloading the motor and cavitation – (Figs. **5.18** and **5.19**).

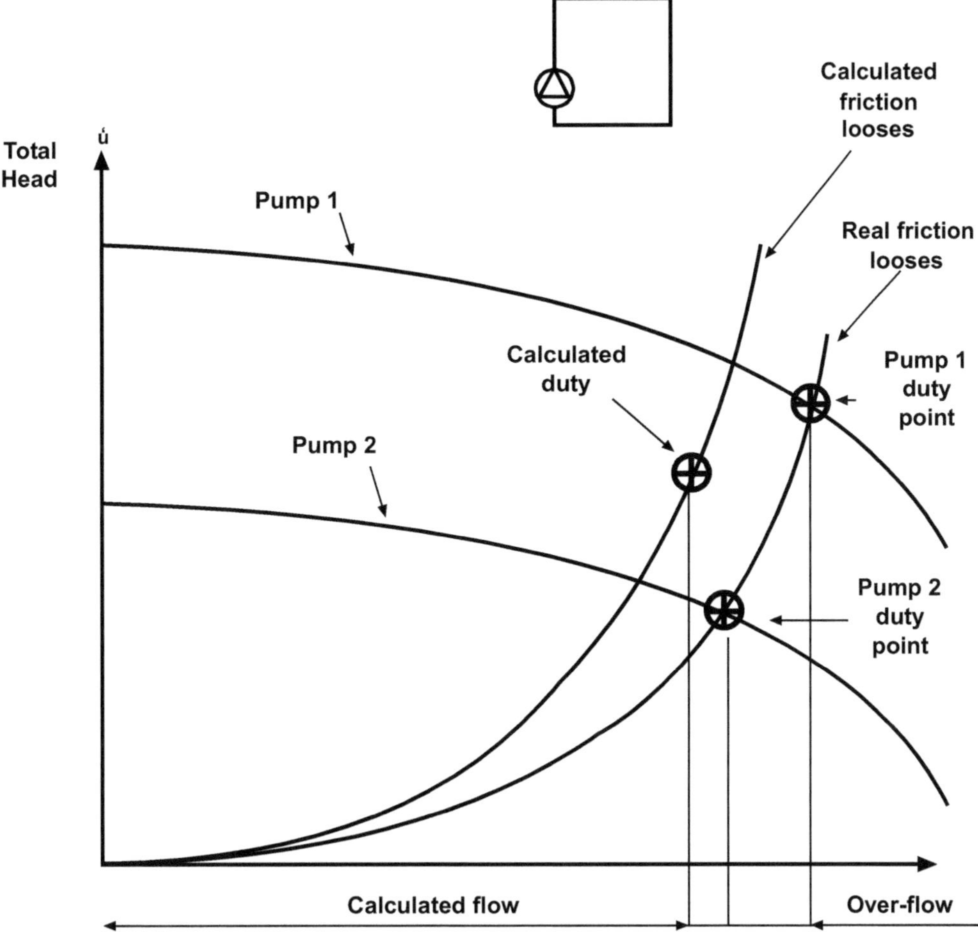

Fig. (**5.18**). Over estimation of friction losses - Closed loop [10].

Risks: Noise in the pipe work; Higher NPSHn requested; More power consumption; Faster wear.

Solutions: Trimming the impeller diameter; Increase the friction losses (diaphragm, valves); Reduce the rotation speed.

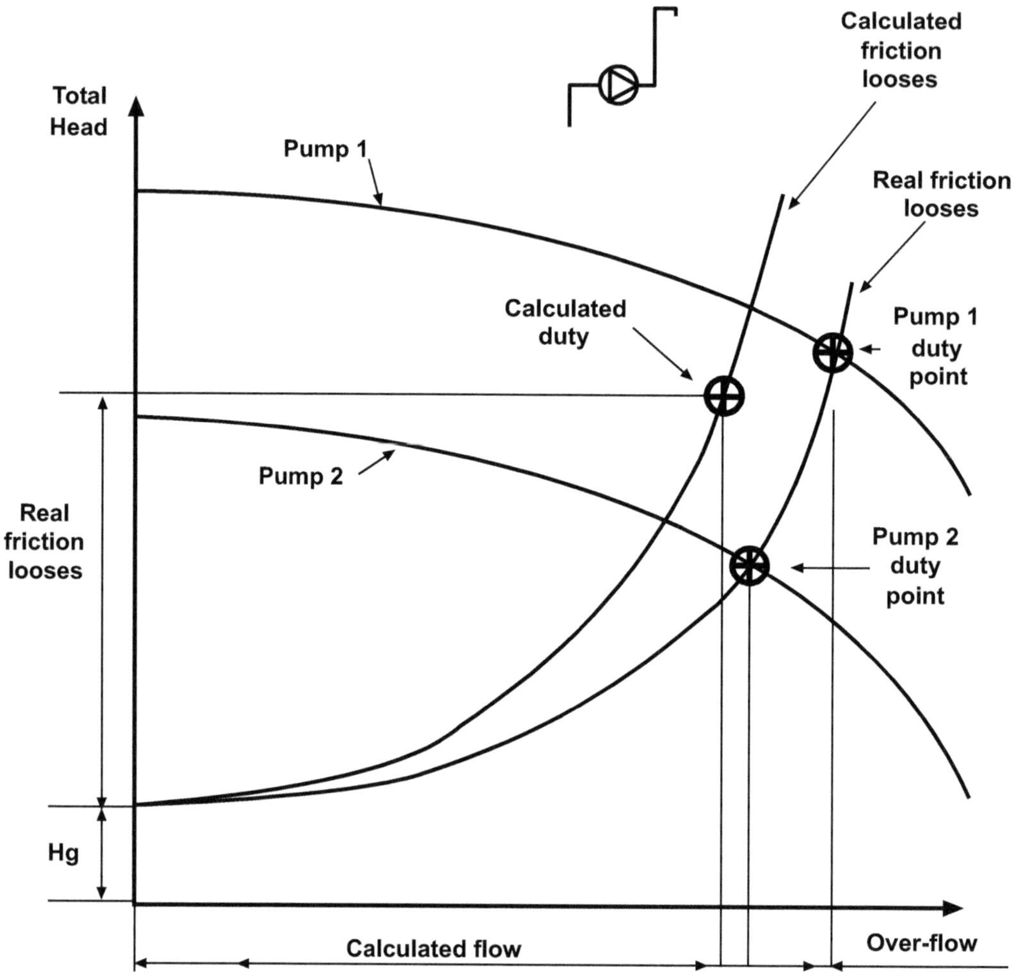

Fig. (5.19). Over estimation of friction losses - Open loop [10].

Risks: Noise in the pipe work; Higher $NPSH_n$ requested; More power consumption**;** Faster wear.

Solutions: Trimming the impeller diameter; Increasing the friction losses (diaphragm, valves)**;** Reducing the rotation speed.

c) Pump's hydraulic characteristics for connected pumps.

Same pumps in series (Fig. **5.20**) – in this case the liquid is passing through the two impellers and therefore $Q_a = Q_{p1} = Q_{p2}$. The head of the assembly is $H_a = H_{p1} + H_{p2}$. The efficiency of the assembly is:

$$\eta_a = \frac{\gamma Q_a H_a}{P_1 + P_2} \quad \textbf{(5.26)}$$

where: P_1, P_2 are the electric powers of the pumps.

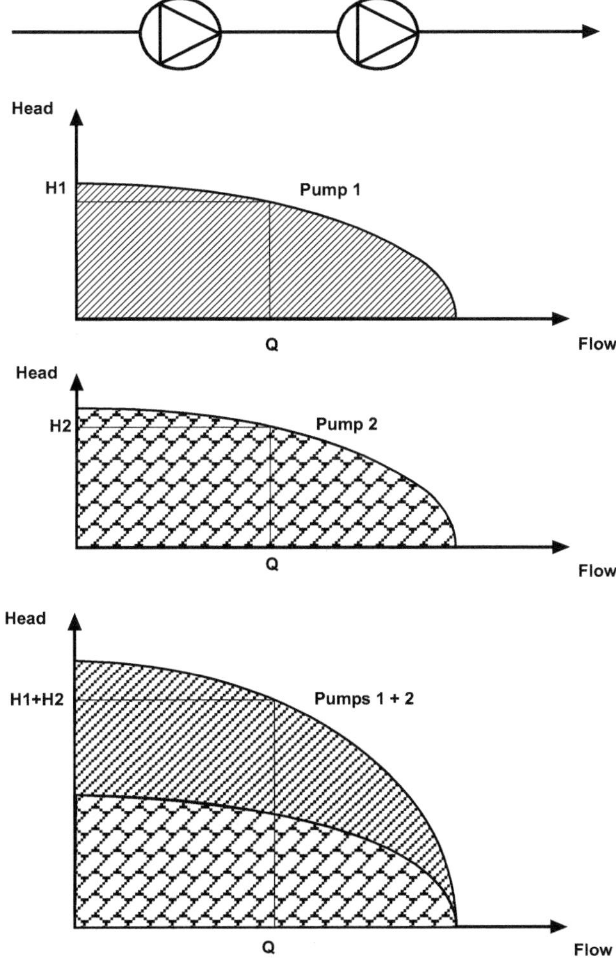

Fig. (5.20). Same pumps in series [10].

Same pumps in parallel (Fig. **5.21**) – in this case the liquid is passing through the two impellers and therefore $Q_a = Q_{p1} + Q_{p2}$ and $H_a = H_{p1} = H_{p2}$.

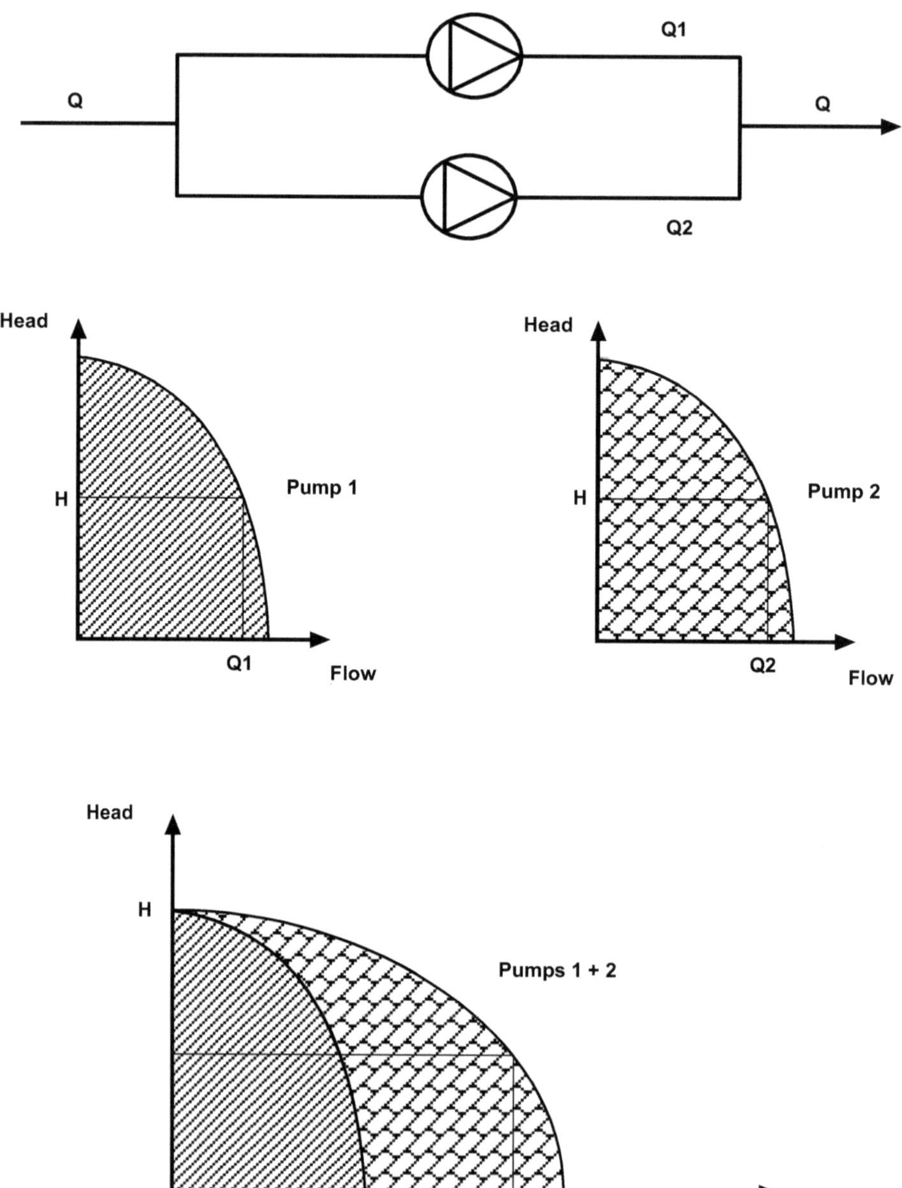

Fig. (5.21). Same pumps in parallel [10].

The efficiency of the assembly is:

$$\eta_a = \frac{\gamma Q_a H_a}{\frac{1}{2}\frac{\gamma Q_c H_c}{\eta_c} + \frac{1}{2}\frac{\gamma Q_c H_c}{\eta_c}} = \eta \quad \textbf{(5.27)}$$

i.e. the value for the assembly efficiency is equal to one of the two pumps.

For different pumps connected in series or parallel same algorithms are valid – (Figs. **5.22** and **5.23**).

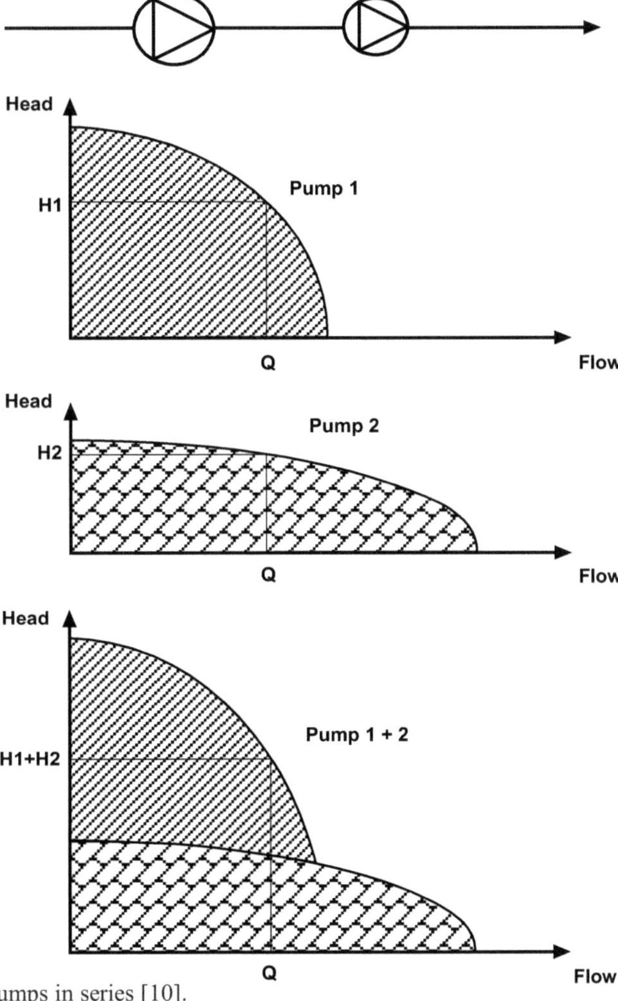

Fig. (5.22). Different pumps in series [10].

If the pumps are not of the same size the efficiency of the assembly is:

$$\eta_a = \frac{\gamma Q_a H_a}{\frac{\gamma Q_1 H_1}{\eta_1} + \frac{\gamma Q_2 H_2}{\eta_2}} = \frac{H_a}{\frac{H_1}{\eta_1} + \frac{H_2}{\eta_2}} \tag{5.28}$$

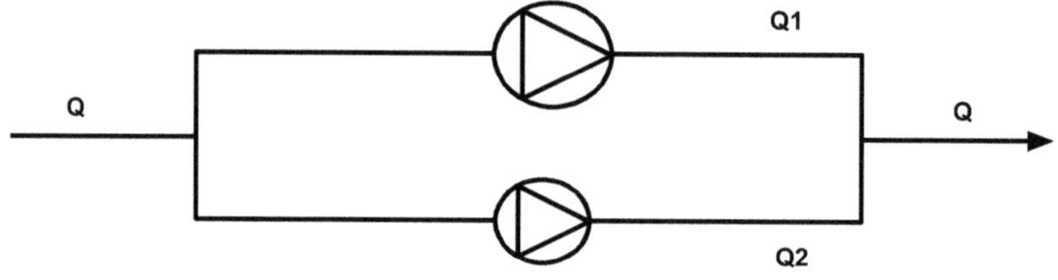

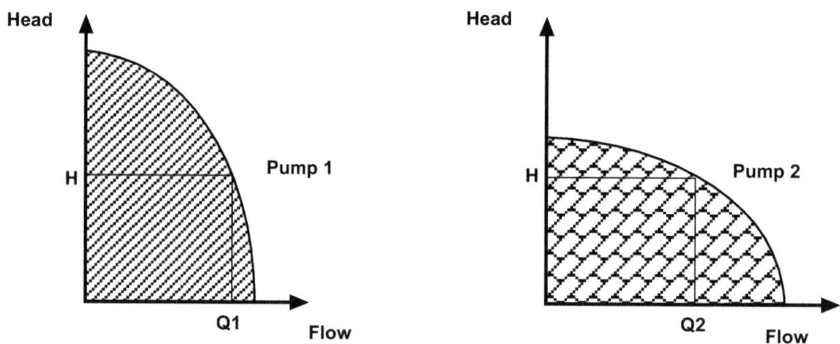

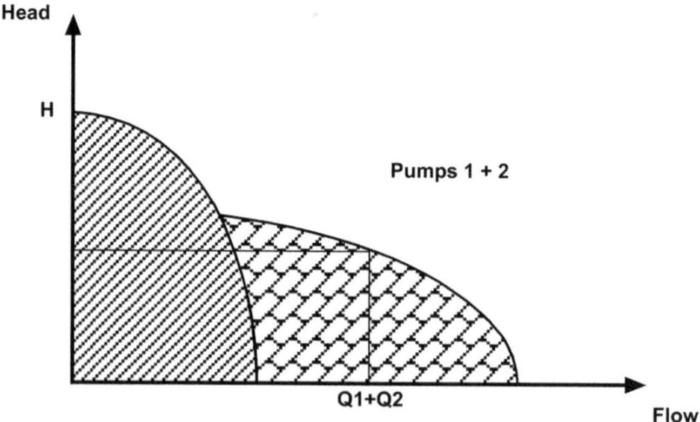

Fig. (5.23). Different pumps in parallel [10].

If the pumps are not of the same size the efficiency of the assembly is:

$$\eta_a = \frac{\gamma Q_a H_a}{\frac{\gamma Q_1 H_1}{\eta_1} + \frac{\gamma Q_2 H_2}{\eta_2}} = \frac{Q_a}{\frac{Q_1}{\eta_1} + \frac{Q_2}{\eta_2}} \tag{5.29}$$

5.12. PUMP SELECTION AND PERFORMANCE CHARTS

The data required for selecting a pump size *i.e.* the flow rate and the head (Q, H) of the desired operating point are assumed to be known from the system requirements - from the application. The electric main frequency is also given. The number of stages can be settled out by using available charts – (Figs. **5.24** and **5.25**). Also further details of the chosen pump as regarding NPSH$_r$ can be obtained. The complete characteristics curves of a centrifugal pump shown in Fig. (**5.26**).

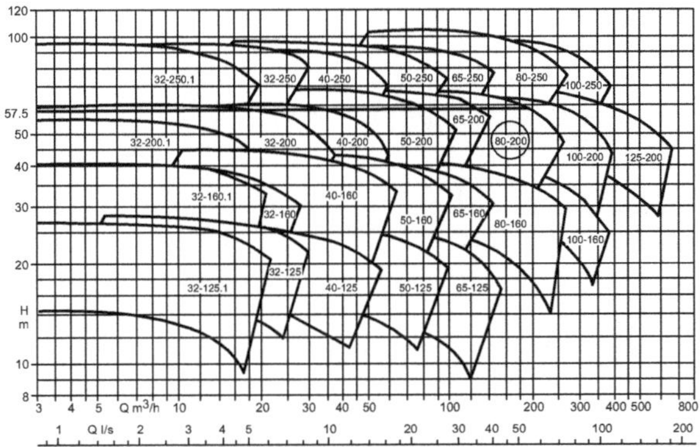

Fig. (5.24). Selection chart for a volute casing pump series for n=2900 rpm
(First number - nominal diameter of the discharge nozzle; second number - nominal impeller diameter) [10].

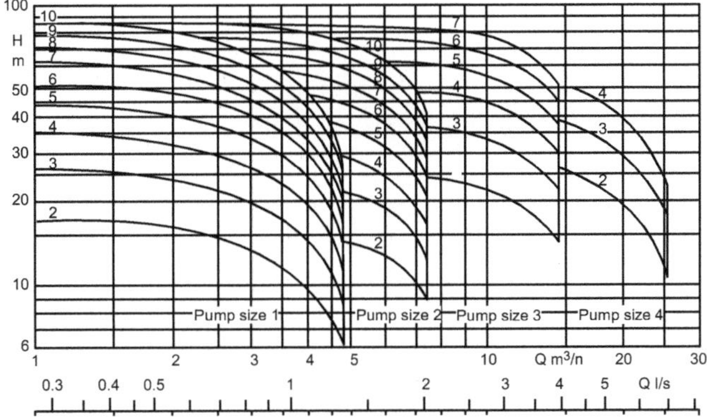

Fig. (5.25). Selection chart for a multistage pump series for N = 2900 rpm [6].

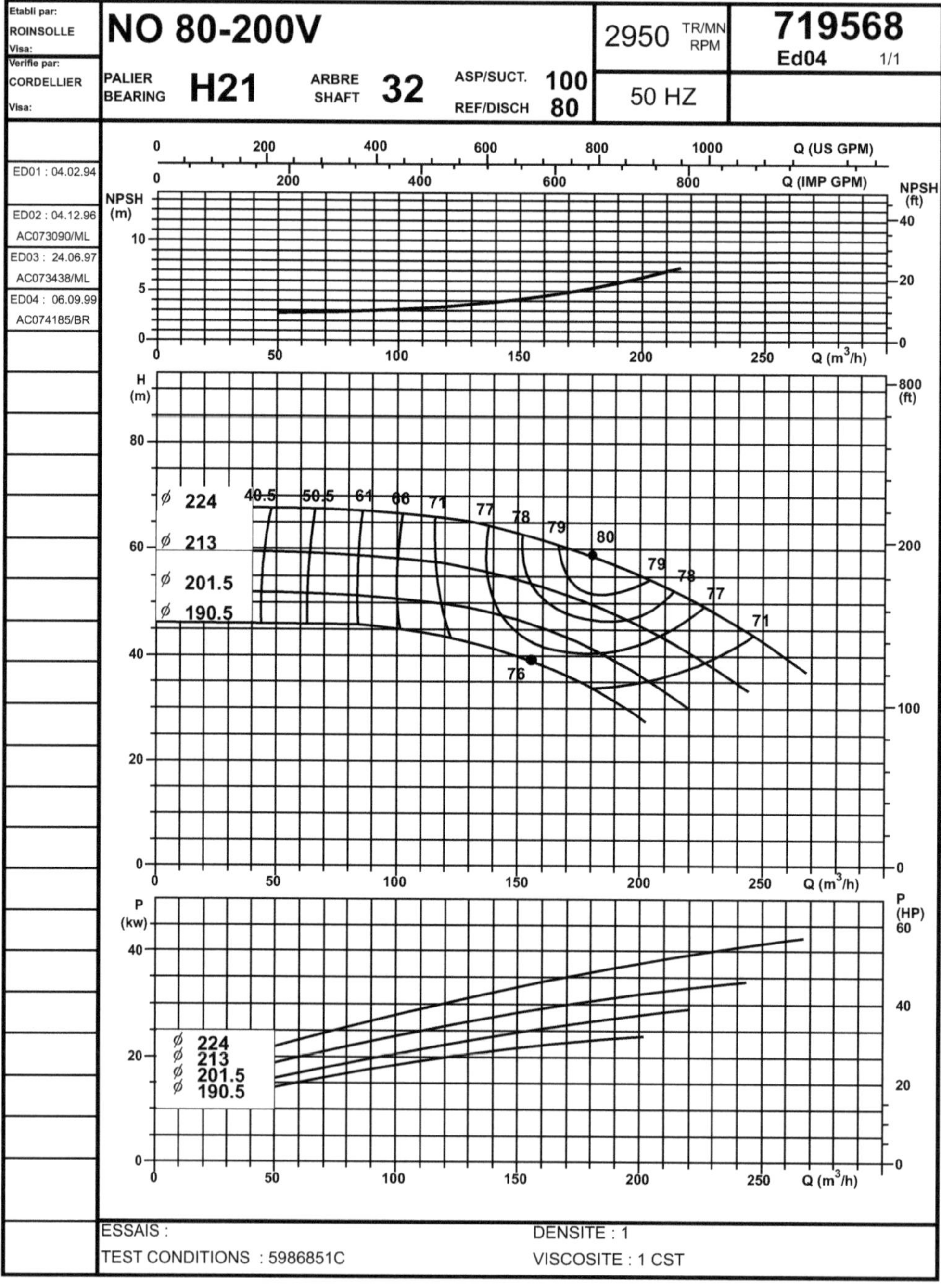

Fig. (5.26). Characteristics curves of centrifugal pumps [10].

5.13. PUMP AS TURBINE

When water enters the rotor at the periphery and flows inward the machine runner it acts as a turbine. When water enters at the center of impeller and flowing outward, the machine acts as a pump. This machine is connected to a motor/generator and is used at pumped storage hydroelectric plants. The pump turbine aggregate is pumping water from a lower reservoir to an upper reservoir during off-peak load periods and is working as a hydraulic turbine during the peak power generation needs.

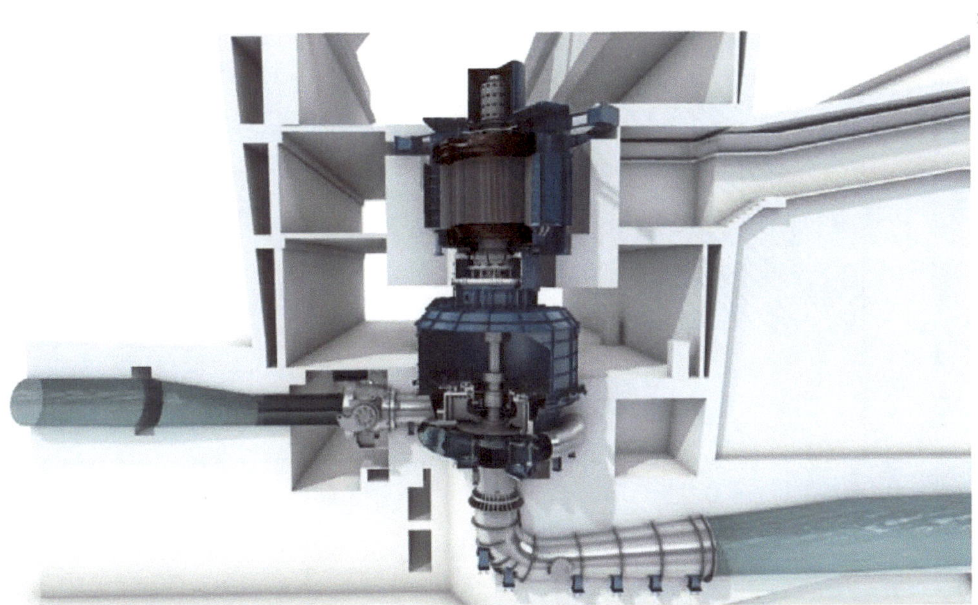

Fig. (5.27). Pumped-Storage Power Plant Hidden Deep in the Swiss Alps – 1,450-MW [12].

The main challenge when using a pump as turbine (PAT) is to predict performance of the machine because unlike a turbine, a centrifugal pump does not have any adjustable guide vanes. If the running speed is fixed, a PAT is able to perform efficiently only for one set of head and flow values. Therefore it is essential to select a pump that correlates with the requested head and discharge availabe at the planned site.

5.13.1. Pump Turbine Classification and Selection

The application of PAT is more suitable for radial and mixed flow serving low specific speed turbine as shown in Fig. (**5.28**). Axial flow pumps are exposed to cavitation, more expensive and thus have a reduced cost-benefit factor.

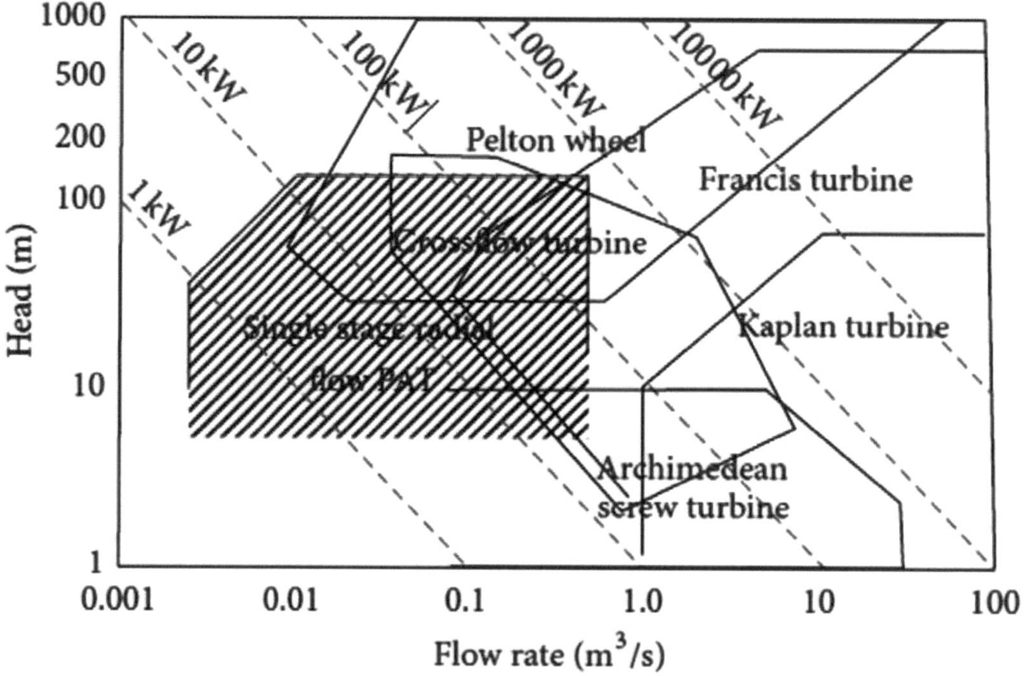

Fig. (5.28). Range of application for different PAT for hydropower [11].

In order to select a pump that will work as turbine it is important to obtain the relation between flow rate and head of both machines based on the characteristics curves - (Fig. **5.29**) and experimental data.

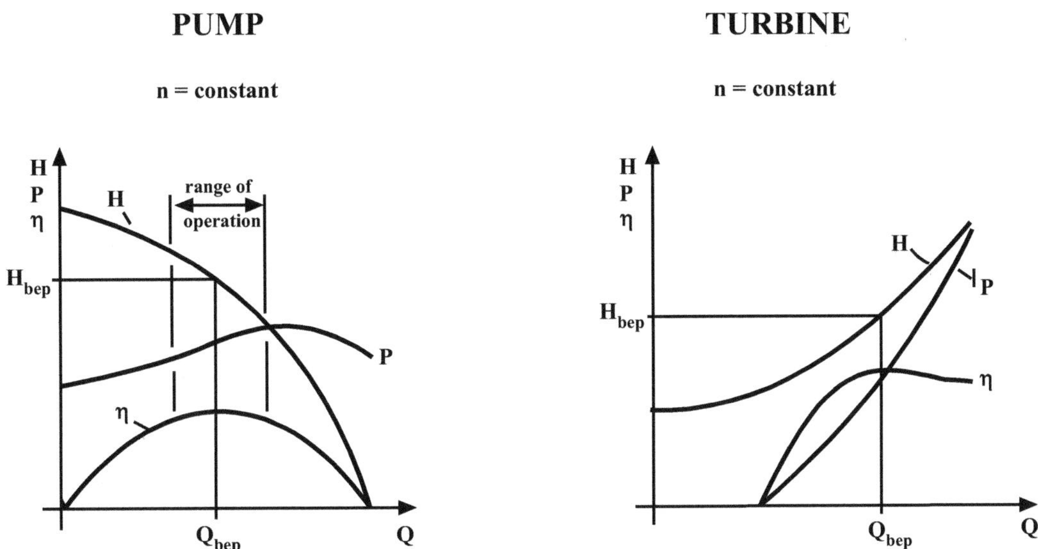

Fig. (5.29). Typical performance curves of pumps and turbines, [11].

If circulation losses because the relative vortex in the pump and turbine rotors are neglected, the relation between the ideal head H_E and the pump and turbine heads are as following:

$$H_p = H_E \eta_p \; ; \; H_t = \frac{H_E}{\eta_t} \Rightarrow \frac{H_p}{H_t} = \eta_p \eta_t \tag{5.30}$$

Experimental data highlighted that efficiency in turbine mode can reach the values of the pump mode (at best efficiency point BEP). Therefore, considering an efficiency of 80%, one obtaines orientative value for the ratio between turbine and pump head [13]:

$$\frac{H_p}{H_t} = {\eta_p}^2 = 0.64 \; ; \; \frac{H_t}{H_p} = 1.56 \tag{5.31}$$

According to the relation, pumps operated as turbines require a net head higher that in pump mode which is between 30% and 150% in order to work at their best point efficiency. Therefore, when selecting a pump for a certain site (*i.e.* head and flow conditions) a smaller pump must be considered in turbine mode than for the same conditions in pump mode.

SOLVED PROBLEMS

Q.1- A centrifugal pump lifts water under a static head of 36 m of which 4 m/s is suction lift. Suction lift and delivery pipes are both 150 mm in diameter. The head loss in suction pipe is 1.8 m and in delivery pipe 7 m. The impeller 380 mm in diameter and 25 mm wide at the mouth and revolves at 1200 rpm. Its exit angle is 35°. If the manometric efficiency, of the pump is 82%. Determine the discharge and the pressure at the suction and delivery branches of the pump. The flow at the inlet is radial *i.e.* $\alpha_1 = 90^\circ$.

Solution:-

$$H_{static} = 36\,m \ ;\ D_2 = 380\,mm \ ; H_s = 4\,m$$

$$B_2 = 25\,mm \ ;\ H_d = 36 - 4 = 32\,m \ ; N = 1200\,rpm$$

$$H_{Ls} = 1.8\,m ;\ H_{Ld} = 7\,m \ ;\ d_d = d_s = 150\,mm$$

$$\eta_{mano} = 82\% \ ;\ \beta_2 = 35^\circ$$

Total head to be supplied by the pump

$$H_{mano} = H_{static} + \sum H_L$$

$$= 36 + 1.8 + 7 = 44.8\,m$$

Peripheral velocity of the impeller at outlet

$$U_2 = \frac{\pi D_2 N}{60} = \frac{\pi \times 0.38 \times 1200}{60}$$

$$= 23.9\,m/s$$

Since the flow radial at inlet $\alpha_1 = 90^0$

$$\therefore \eta_{mano} = \frac{H_{mano}}{\frac{V_{u2}U_2}{g}} = \frac{44.8}{\frac{V_{u2} \times 23.9}{9.81}} = 0.82$$

$$\therefore V_{u2} = 22.4\, m/s$$

$$\tan \beta_2 = \frac{V_{f2}}{U_2 - V_{u2}} \quad from\ velocity\ triangle$$

$$= \frac{V_{f2}}{23.9 - 22.4} \quad \therefore V_{f2} = 1.05\, m/s$$

$$= \frac{V_{f2}}{23.9 - 22.4} \quad \therefore V_{f2} = 1.05\, m/s$$

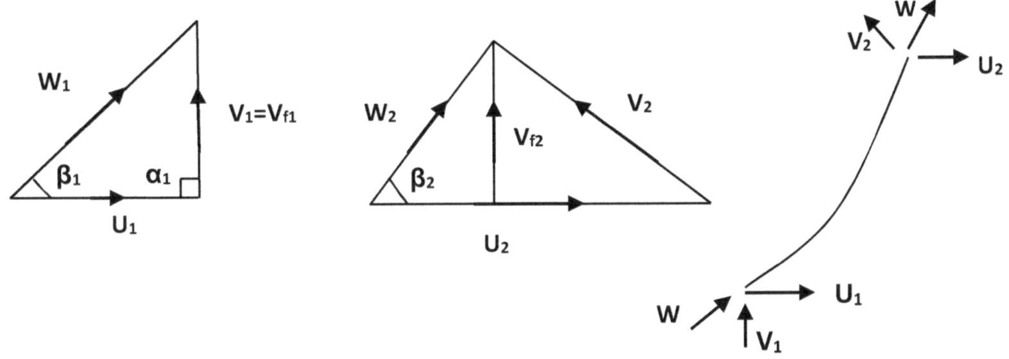

$$and\ Q = \pi D_2 B_2 V_{f2} = \pi \times 0.38 \times 0.025 \times 1.05$$

$$= 0.0314\, m^3/s$$

Velocity in suction or delivery pipe

$$V_d = V_s = \frac{Q}{a_p} = \frac{0.0314}{\frac{\pi}{4} \times (0.15)^2} = 1.78\, m/s$$

$$\text{Velocity head} = \frac{V^2}{2g} = \frac{(1.78)^2}{2 \times 9.81} = 0.161\, m$$

Total pressure head on the delivery

$$side = H_d + H_{Ld} + \frac{{V_d}^2}{2g} = 32 + 7 + 0.161$$

$$= 39.161\, m$$

$$P_d = \gamma H_d = 9.81 \times 39.161 = 384\, kPa.$$

$$Total\ pressure\ head\ on\ the\ suction\ side\ = 4 + 1.8 + 0.161 = 5.961\, m$$

$$P_s = \gamma H_s = 58.5\, kPa.$$

Q.2- A radial single stage, double suction, C.P. is manufactured for the following data:

$$Q = 75\ \frac{l}{s}\ ;\ D_1 = 100\ mm\ ;\ D_2 = 270\ mm$$

$$N = 1750\, rpm\ ;\ B_1 = 25\ mm\ perside\ ;\ B_2 = 23\ mm\ intotal$$

$$\eta_{overall} = 55\%\ ; P_{mech.losse}\ = 1.04\ kW\ :\ \alpha_1 = 90°\ ;\ \beta_2 = 27°$$

Leakage losses = 2.25 l/s

$$contraction\ coefficient\ due\ to\ blde\ thicknees = 0.87$$

Determine: -

(a) The inlet blade angle β_1

(b) The angle at which the water leaves α_2

(c) The speed ratio Ø

(d) The absolute velocity of water leaving Impeller V_2

(e) The manometric efficiency

(f) The volumetric and mechanical efficiency

Solution:-

Total quantity of water handled by the pump

$$Q_t = Q_d + Q_{leakage} = 75 + 2.25 = 77.25\ l/s$$

$$Q_{per\ side} = \frac{77.25}{2} = 38.625\ l/s$$

a) Peripheral speed at outlet $U_1 = \frac{\pi D_1 N}{60}$

$$U_1 = \frac{\pi \times 0.1 \times 17500}{60} = 9.15\ m/s$$

$$\text{Area of flow at inlet} = \pi D_1 B_1 \times 0.87$$

$$= \pi \times 0.025 \times 0.1 \times 0.87 = 0.00683\ m^3$$

$$V_{f1} = \frac{Q}{area\ of\ flow} = \frac{38.625 \times 10^{-3}}{0.00683} = 5.66\ m/s$$

$$\alpha_1 = 90°$$

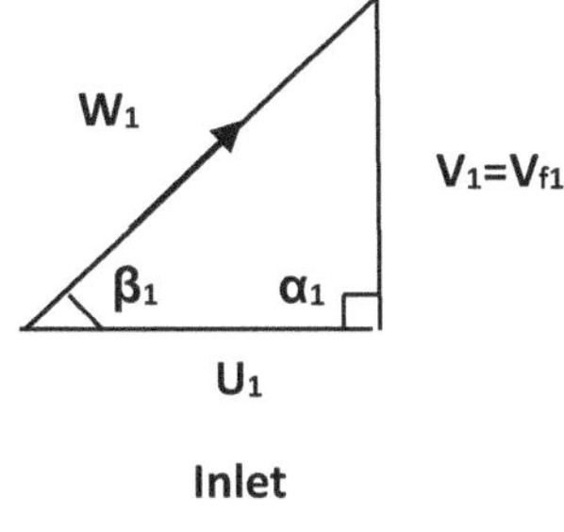

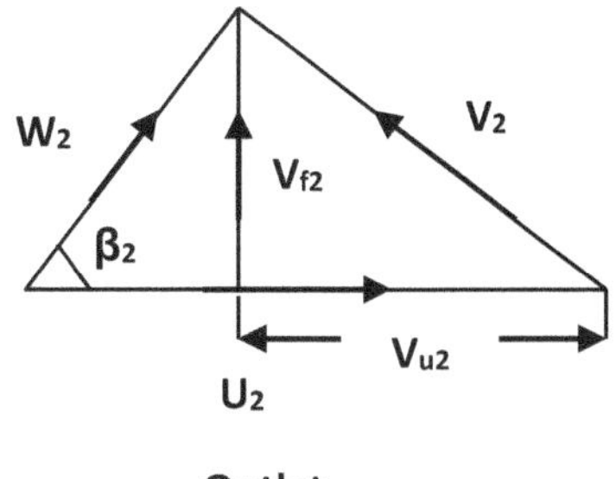

From inlet velocity triangle

$$\tan\beta_1 = \frac{V_{f1}}{U_2} = \frac{5.66}{9.15} = 0.62$$

$$\beta_1 = 31.8°$$

b) $$V_{f2} = \frac{Q}{area\ of\ flow} = \frac{38.625 \times 10^{-3}}{\pi \times 0.29 \times \left(\frac{23}{2} \times 10^{-3}\right) \times 0.87}$$

$$V_{f2} = 4.25\ m/s$$

$$U_2 = \frac{\pi D_2 N}{60} = \frac{\pi \times 0.29 \times 1750}{60}$$

$$= 26.55\ m/s$$

$$Now \quad \beta_2 = 27°$$

From velocity triangle at outlet

$$\tan\beta_2 = \frac{V_{f2}}{U_2 - V_{u2}} = \frac{4.25}{26.55 - V_{u2}}$$

$$\therefore\ V_{u2} = 18.21\ m/s$$

$$or \quad \tan\alpha_2 = \frac{V_{f2}}{V_{u2}} = \frac{4.25}{18.21}$$

$$\alpha_2 = 13.13°$$

c) $$Speed\ ratio \quad \emptyset = \frac{U_2}{\sqrt{2gH_{mano}}} = \frac{26.55}{\sqrt{2 \times 9.81 \times 30}}$$

$$\emptyset = 1.0905$$

d) Absolute velocity of water leaving the impeller

$$V_2 = \frac{V_{u2}}{\cos \alpha_2} = \frac{18.21}{0.971} = 18.75\ m/s$$

$$e)\ Manometric\ efficiency\ \eta_{mano} = \frac{gH_{mano}}{V_{u2}U_2}$$

$$\eta_{mano} = \frac{9.81 \times 30}{26.55 \times 18.21} = 61\%$$

$$Volumetric\ efficiency\ \eta_Q = \frac{Q}{Q_{total}} = \frac{75}{77.25} = 97.1\%$$

$$Shaft\ power\ SKW = \frac{\gamma QH}{\eta_{overall}} = \frac{9.81 \times 75 \times 10^{-30} \times 30}{0.55}$$

$$= 20.65\ kW$$

$$\eta_{mech} = \frac{SKW - P_{mech\ loss}}{SKW} = \frac{20.65 - 1.04}{20.65} = 94.9\%$$

A six stage C.P. delivers 120 l/s against a head pressure rise of 5000 kN/m^3. Determine its specific speed if it rotates at 1450 rpm.

Solution:-

$$N_s = \frac{N\sqrt{Q}}{H^{3/4}}$$

$$N_s = \frac{1450\sqrt{0.12}}{\left(\frac{5000}{9.81 \times 6}\right)^{3/4}} = 17.9 \approx 18$$

Q.4- A C.P. heaving an overall efficiency of 75% delivers 1220 l/min to a height of 12 m through a pipe of 100 m diameter and 90 m length. If f=0.0048, calculate the power required to drive the pump.

Solution:-

$$Q = \frac{1.820}{60} = 0.0303\ m^3/s$$

$$\eta_o = 0.75\ ;\ H_{st} = 18\ m\ ; d = 0.1\ m\ ; l = 90\ m$$

$$Friction\ head\ loss\ h_f = f.\frac{L}{D}.\frac{V^2}{2g}$$

$$V = \frac{Q}{A} = \frac{0.0303}{\frac{\pi}{4}(0.1)^2} = 3.86\ m/s$$

$$\therefore h_f = 0.0048 \times \frac{90}{0.1}.\frac{(3.86)^2}{2 \times 9.81} = 3.28\ m$$

$$Total\ head\ H_{mano} = H_{st} + H_f = 18 + 3.28$$

$$= 21.28\ m$$

$$P = \frac{\gamma Q H}{\eta_o} = \frac{9.81 \times 0.0303 \times 21.28}{0.75}$$

$$= 8.43\ kW$$

Q.5- Calculate the vane angle at inlet of a C.P. impeller having 300 mm diameter at inlet and 600 mm diameter at outlet. The outlet angle of the vane is 45° and the entry of the pump is radial, the pump runs at 1000 rpm and the velocity of flow through the impeller is constant at 3 m/s. Also calculate the power per unit mass developed by the pump and the velocity and direction of water at outlet.

$$\alpha_1 = 90°\ ;\ \beta_2 = 45°\ ;\ N = 1000\ rpm$$

$$D_1 = 300\ mm\ ; D_2 = 600\ mm\ ;\ V_{f1} = V_{f2} = 3\ m/s$$

$$Find\ \beta_1\ , V_2\ , \alpha_2?$$

Solution:-

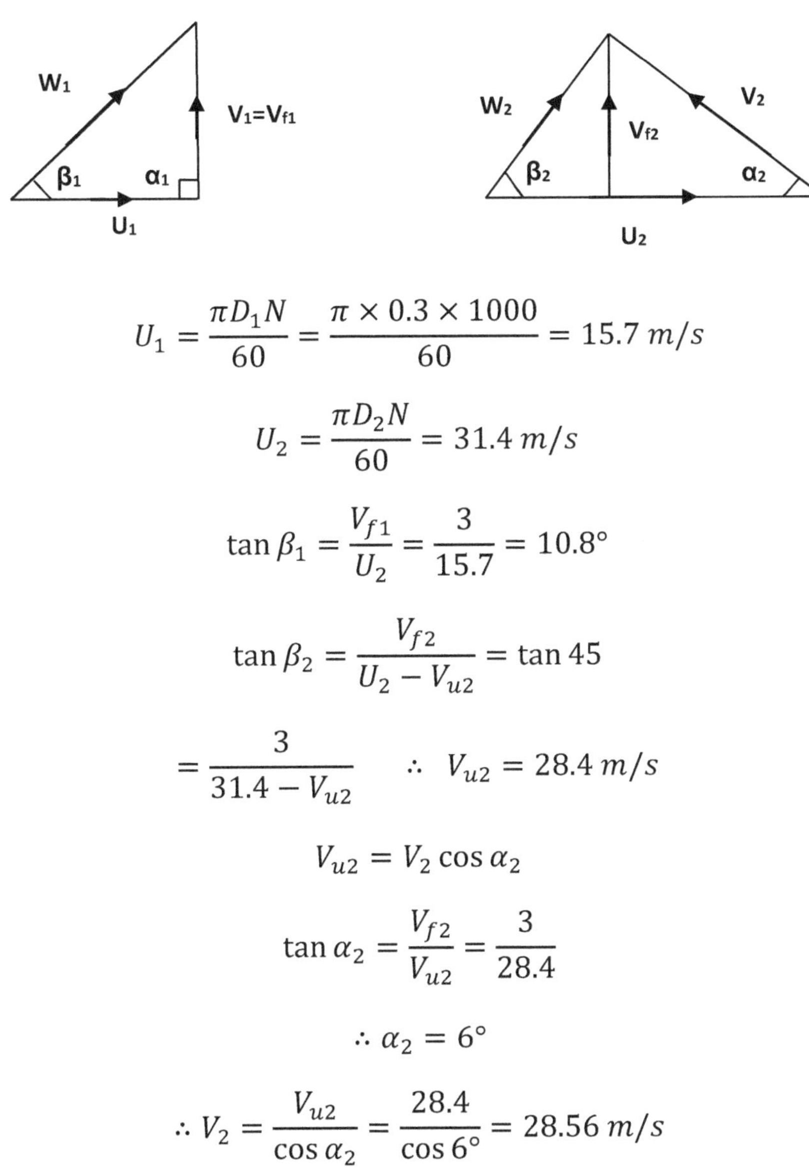

$$U_1 = \frac{\pi D_1 N}{60} = \frac{\pi \times 0.3 \times 1000}{60} = 15.7\ m/s$$

$$U_2 = \frac{\pi D_2 N}{60} = 31.4\ m/s$$

$$\tan\beta_1 = \frac{V_{f1}}{U_2} = \frac{3}{15.7} = 10.8°$$

$$\tan\beta_2 = \frac{V_{f2}}{U_2 - V_{u2}} = \tan 45$$

$$= \frac{3}{31.4 - V_{u2}} \qquad \therefore\ V_{u2} = 28.4\ m/s$$

$$V_{u2} = V_2 \cos\alpha_2$$

$$\tan\alpha_2 = \frac{V_{f2}}{V_{u2}} = \frac{3}{28.4}$$

$$\therefore \alpha_2 = 6°$$

$$\therefore V_2 = \frac{V_{u2}}{\cos\alpha_2} = \frac{28.4}{\cos 6°} = 28.56\ m/s$$

$$or\ \ V_2 = \sqrt{{V_{f2}}^2 + {V_{u2}}^2} = 28.56\ m/s$$

Q.6- A C.P. delivers 30 l/s of water against a head of 12 m and running at 1450 rpm. Requires 4.5 kW. Determine the discharge, head of the pump and power required, if the pump runs at 1800 rpm for the same size.

Solution:-

$$N = 1450\, rpm\ ; Q = 30\frac{l}{s}\ ; H = 12\, m\ ;\ Power = 4.5\, kW$$

$$Find\ H\ , Q\ , P\ \ for\ N = 1800\, rpm D_1 = D_2 \quad same\ size$$

$$N_{sm} = N_{sp}$$

$$\frac{N_m}{N_p} = \frac{D_p}{D_m}\sqrt{\frac{H_m}{H_p}}$$

$$\frac{1450}{1800} = 1 \times \sqrt{\frac{12}{H_p}} \qquad \therefore\ H_p = 18.5\, m$$

Also

$$\frac{Q_m}{Q_p} = \left(\frac{D_m}{D_p}\right)^2 \sqrt{\frac{H_m}{H_p}}$$

$$\frac{30}{Q_p} = 1 \times \sqrt{\frac{12}{18.5}} \qquad Q_p = 37.3\, l/s$$

$$\frac{P_m}{P_p} = \left(\frac{D_m}{D_p}\right)^2 \times \left(\frac{H_m}{H_p}\right)^{3/2}$$

$$\frac{4.5}{P_p} = 1 \times \left(\frac{12}{18.5}\right)^{3/2}$$

$$P_p = 8.6\, kW$$

Q.7- A C.P. impeller runs at 950 rpm. Its external and internal diameters are 500 mm and 250 mm. The blade is set back at an angle of 35° to the water rim. If the radial velocity of water through the impeller be maintained constant at 2 m/s. Find the angle of the blade at inlet, the velocity and direction of water at outlet and the work done by the impeller per kN of water.

Solution:-

$$\beta_2 = 35° \ ; N = 950\, rpm$$

$$D_1 = 0.25\, m \ ; D_2 = 0.5\, m \ ; \ V_{f1} = V_{f2} = 2\, m/s$$

$$U_2 = \frac{\pi D_2 N}{60} = \frac{\pi \times 0.5 \times 950}{60} = 24.87\, m/s$$

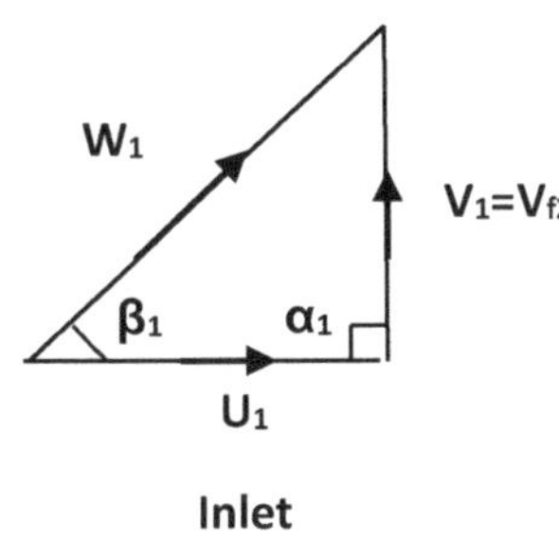

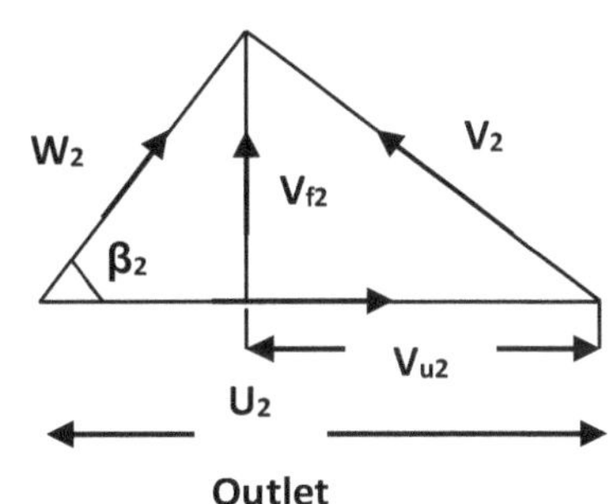

(i) $U_1 = \frac{U_2}{2} = 12.435\, m/s$ $\alpha_1 = 90°$

$$\therefore V_{u1} = 0 \quad radial\, flow$$

$$\tan\beta_1 = \frac{V_{f1}}{U_1} = \frac{2}{12.435} \quad \therefore \ \beta_1 = 9.146°$$

(ii) $$V_{u2} = U_2 - \frac{V_{f2}}{ta \quad 2} = 24.87 - \frac{2}{\tan 35}$$

$$V_{u2} = 22.1\, m/s$$

$$V_2 = \sqrt{V_{f2}{}^2 + V_{u2}{}^2} = \sqrt{2^2 + (22.1)^2} = 22.2\, m/s$$

(iii) $\alpha_2 = ?$

$$\tan\alpha_2 = \frac{V_{f2}}{V_{u2}} = \frac{2}{22.1} = 0.09$$

$$\therefore\ \alpha_2 = 5.142°$$

(iv) $$Work\ done\ per\ unit\ weight = \frac{V_{u2}U_2}{g}$$

$$= \frac{22.1 \times 24.87}{9.81} = 56\ kN$$

Q.8- A C.P. of 1.2 m diameter. Runs at 200 rpm and pumps 1880 l/s, the average lift being 6 m. The angle which the blade make at exit with the tangent to the impeller is 26°. And the radial velocity of flow is 2.5 m/s. Determine the useful power and the efficiency.

Solution:-

$$D_2 = 1.2\ m;\ N = 200\ rpm;\ Q = 1.88\ m^3/s$$

$$H = 6\ m;\ \beta_2 = 26°;\ V_{f2} = V_{f1} = 2.5\ m/s$$

$$D_1 = 0.6\ m$$

$(i)\quad P = \gamma QH = 9.81 \times 1000 \times 1.88 \times 6 = 110\ kW$

$(ii)\quad Theoretical\ head = \frac{V_{u2}U_2}{g}$

To manometric head (H_{mano}), if losses is neglected $\alpha_1 = 90°$

$$\therefore\ H_{mano} = \frac{V_{u2}U_2}{g}$$

$$U_2 = \frac{\pi D_2 N}{60} = \frac{\pi \times 1.2 \times 200}{60} = 12.56\ m/s$$

$$V_{u2} = U_2 - \frac{V_{f2}}{\tan \beta_2} = 12.56 - \frac{2.5}{\tan 26°}$$

$$= 7.44\ m/s$$

$$\therefore\ H_{mano} = \frac{7.44 \times 12.56}{9.81} = 9.526\ m$$

$$\therefore\ \eta_{mano} = \frac{6}{9.526} \times 100 = 63\%$$

5.14. POSITIVE DISPLACEMENT PUMPS

A positive displacement pump is a hydraulic machinery in which an amount of volume contained in a certain cavity is forced to move from inlet section to discharge section by means of a piston. Positive displacement pumps can be classified as reciprocating (piston, diaphragm) or rotary type (single rotor, multiple rotor).

5.14.1. Reciprocating Pumps Classification

Reciprocating pumps may be classified as follows:

a) According to acting of water

1. single acting pump.
2. double acting pump.

b) According to number of cylinders

1. single cylinder pump.
2. double cylinder pump and

c) According to existence of air vessels

1. with air vessel, and
2. without air vessel

The advantages of centrifugal pump over the reciprocating type can be summarized in the following Table **5.3**.

Table 5.3. A comparison between a centrifugal pump and a reciprocating pumps.

S.No.	Centrifugal Pump	Reciprocating Pump
1.	Simple in construction, because of less number of parts.	Complicated in construction because of more number of parts.
2.	Total weight of the pump is less for a given discharge.	Total weight of the pump is more for a given discharge.
3.	Suitable for large discharge and Smaller heads.	Suitable for less discharge and higher heads.
4.	Requires less floor area and Simple foundation.	Requires more floor area and Comparatively heavy foundation.
5.	Less wear and tear.	More wear and tear.
6.	Maintenance cost is less.	Maintenance cost is More.
7.	Can handle dirty water.	Cannot handle dirty water.
8.	Can run at higher speeds.	Cannot run at higher speeds.
9.	Its delivery is continuous.	Its delivery is pulsating.
10.	No air vessels are required.	Air vessels are required.
11.	Thrust on the crankshaft is uniform.	Thrust on the crankshaft is not uniform. Much care is required in Operation.
12.	Operation is quite simple.	Does not need priming.
13.	Needs priming.	It has more efficiency.
14.	It has less efficiency.	

5.14.2. Theory of Reciprocating Pumps

Reciprocating pumps can be studied if the kinetics of the mechanism that is responsible for piston movement is related with hydraulics of the liquid volume trapped between the cylinder and piston. In order derive the basic relations describing how the pump operate, basic structure and geometry data are considered as shown in Fig. (**5.30**).

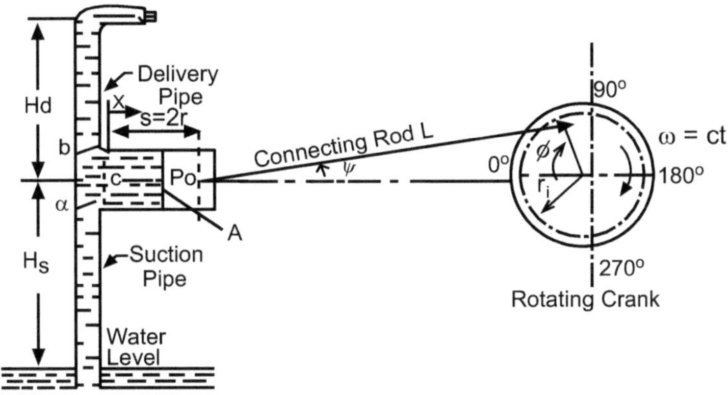

Fig. (5.30). Parts of a reciprocating pump [4].

where:

- C is the cylinder in which piston P works. The motion of the piston is obtained by a connecting rod, which connects the piston and the rotating crank;
- suction pipe, connecting the source of water and the cylinder;
- delivery pipe, into which the water is discharged from the cylinder;
- a valve *a*, which admits the flow from the suction pipe into the cylinder;
- a valve *b*, which admits the flow from the cylinder into the delivery pipe.

Considering the geometric data from Fig. (**5.30**) it can be computed the piston displacing x:

$$x = r_i + L - r_i\cos\varphi - L\cos\psi$$

$$\frac{L}{\sin\varphi} = \frac{r_i}{\sin\psi}; \qquad (5.32)$$

$$\cos\psi = \sqrt{1 - \lambda^2 \sin\varphi}$$

therefore:

$$x = r_i\left(1 - \cos\varphi + \frac{1 - \sqrt{1 - \lambda^2\sin^2\varphi}}{\lambda}\right)$$

where: $\lambda = r_i/L$ be the coefficient of the crank and connecting rod mechanism (generally $\lambda = 1/5;\ 1/6;\ 1/10$).

By differentiation one obtain the velocity *c* and acceleration *j* of the piston pump as follows:

$$c = \dot{x} = r_i\omega\left(\sin\varphi + \frac{\lambda\sin\varphi\cos\varphi}{\sqrt{1-\lambda^2\sin^2\varphi}}\right)$$

$$j = \dot{c} = \ddot{x} = r_i\omega^2\left(\cos\varphi + \lambda\frac{\cos^2\varphi - \sin^2\varphi + \lambda^2\sin^4\varphi}{(1-\lambda^2\sin^2\varphi)^{3/2}}\right)$$

If the length of the connecting rod L is much higher than the crank r_i *i.e.* $L >> r_i$, therefore $\lambda \rightarrow \infty$ and the relations are simpler. Still such a simplification is to be used only for simple pumps *i.e.* one piston. Otherwise the flow irregularities will be unacceptable when computing the gap between the theoretic and effective discharge flow.

5.14.3. Characteristics of Reciprocating Pumps

The characteristic of a reciprocating pump is a diagram showing the pressure head in the cylinder during the suction and delivery strokes, as presented in Fig. (**5.31**):

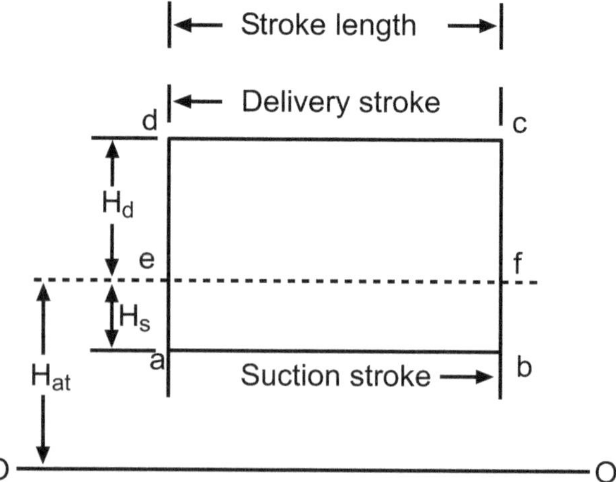

Fig. (5.31). Ideal indicator diagram [14].

where: H_{at} – atmospheric height; H_s – Height of the cylinder axis above the sump water level; H_d – Height of the delivery outlet above the axis of the cylinder. The total height of which liquid is lifted is $H_s + H_d$.

As the liquid flows through the suction and delivery pipes head losses are computed and ideal diagram is changed as presented in Fig. (**5.32**).

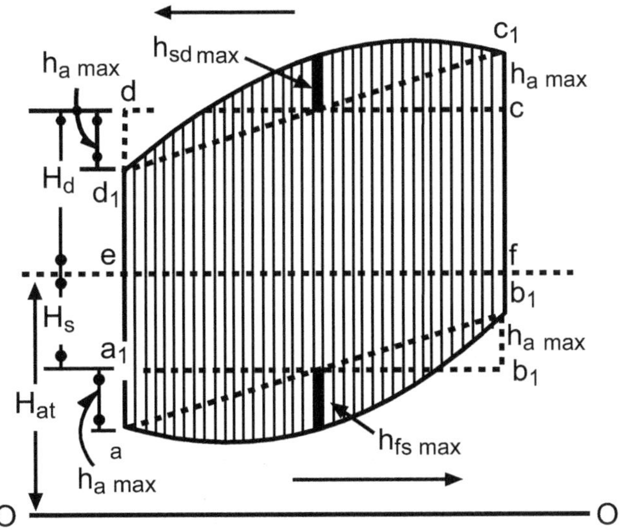

Fig. (5.32). Indicator diagram corrected [14].

In this diagram, acceleration of liquid column in the pipes and the frictions are considered. As frictions has a parabolic variation ($h_{fsuction}$ and $h_{fdischarge}$) due to the velocity *c* of the crank and connecting rod mechanism (as presented above) the loss of head varies from zero at the begining and at the end of the stroke with a maximum value at the middle of the cycle:

$$h_f = \frac{v^2}{2g}\frac{fl}{d} \tag{5.33}$$

where: v = cA/a – the velocity of the liquid in the hypothesis that the piston velocity is equal with the velocity of the liquid (the liquid is always in contact with the piston); a = area of the pipes; f – the coefficient of the pipe; l – the length of the pipe; d – the pipe diameter. Therefore, for the indicator diagram the relations for total pressure head in the cylinder are as follows:

For the suction stroke

a) for the beginning of the suction stroke - $H_{a1} = H_{at} - H_s - h_{amax}$;
b) for the middle of the suction stroke - $H_{1/2} = H_{at} - H_s - h_{fsmax}$;
c) for the end of the stroke – $H_{b1} = H_{at} - H_s + h_{amax}$;

For the delivery stroke

a) for the beginning of the suction stroke – $H_{c1} = H_{at} + H_d + h_{amax}$;
b) for the middle of the suction stroke - $H_{1/2} = H_{at} + H_d - h_{fdmax}$;
c) for the end of the stroke – $H_b = H_{at} + H_d - h_{amax}$;

Power of the pump can be computed as follows:

$$P_{out} = \gamma QH \qquad kW \tag{5.34}$$

$$H = H_s + H_d$$

$H_s = section\ head\ (m)$

$H_d = delivery\ head\ (m)$

The discharge of a single acting pump can be computed as follows:

$$Q = \frac{LAN}{60}\ (m^3/s) \tag{5.35}$$

L: *length of the stroke or piston.*

A: *cross – sectional area of the piston.*

N: *No. of revolutions, per minute of the crank.*

For double acting pump:

$$Q = \frac{2LAN}{60}$$

$Slip\ of\ the\ pump = Q_{th} - Q_a$

$$Percentage\ of\ the\ slip = \frac{Q_{th} - Q_a}{Q_{th}}$$

$$Discharge\ coefficient\ C_d = \frac{Q_a}{Q_{th}}$$

Ex.1: A single acting reciprocating pump has a plunger of diameter 30 cm and stroke of 20 cm. If the speed of the pump is 30 rpm and it delivers 6.5 l/s of water. Find the coefficient of discharge and the percentage slip of the pump.

Solution:-

$$D = 30\ cm \quad \therefore\ A = \frac{\pi}{4}D^2 = 706.86\ cm^2$$

$$L = 20\ cm\ ; N = 30\ rpm$$

$$Q_a = 6.5\ l/s = 6500\ cm^3/s$$

Then

$$Q_{th} = \frac{LAN}{60} = \frac{20 \times 706.86 \times 30}{60}$$

$$= 7068.6\ cm^3/s$$

$$\therefore\ C_d = \frac{Q_a}{Q_{th}} = \frac{6500}{7068.6} = 0.92$$

$$Percentage\ of\ the\ slip = \frac{Q_{th} - Q_a}{Q_{th}}$$

$$= \frac{7068.6 - 6500}{7068.6}$$

$$= 8.04\%$$

Ex.2: A single acting R.P. having a bore of 150 mm diameter and a stroke of 300 mm length discharge 200 l/min at 40 rpm. Neglecting losses find:

1- Theoretical discharge.

2- Coefficient of discharge.

3- Slip of the pump.

4- The power of the pump for H_d =26, H_s =4 m.

Solution:-

$$D = 150\ cm \qquad \therefore A = \frac{\pi}{4} D^2 = 706.86\ cm^2$$

(1)

$$Q_{th} = \frac{LAN}{60} = \frac{0.3 \times 0.0177 \times 40}{60} = 0.00354\ cm^3/s$$

$$= 0.00354\ cm^3/min$$

(2)

$$C_d = \frac{Q_a}{Q_{th}} = \frac{200}{212} = 0.94$$

$$Slip\ of\ the\ pump = Q_{th} - Q_a = 12\ l/min$$

(3)

$$slip\ of\ thePercentage = \frac{Q_{th} - Q_a}{Q_{th}} = 5.66\%$$

(4)

$$The\ Power\ of\ the\ pump\ \text{(theoretical)} = \gamma Q H$$

$$= 9.81 \times 0.00354 \times 30 = 883\ watt$$

$$P_{therotica} = 1.042\ kW$$

$$P_{actual} = 9.81 \times \frac{0.200}{60} \times 30 = 0.981\ kW$$

5.14.4. Air Vessel

Flow rate oscillations at reciprocating pump discharge are due to finite number of pistons. Air vessels are closed chambers fitted on the suction as well as on the delivery side, near the pump cylinder to reduce the accelerating head as shown in Fig. (**5.33**). For low number of pistons, but not only for this case, air vessels mounted on suction and delivery pipes, provides uniform discharge rate.

Function:

a- Suction Side:

(1) Reduces the possibility of flow separation.

(2) Pump can be operated at a higher speed, avoiding cavitation.

(3) Length of the suction pipe below can be increased.

b- Delivery Side:

(1) A large amount of power consumed in supplying accelerating head can be saved.

(2) Constant rate of discharge can be ensured.

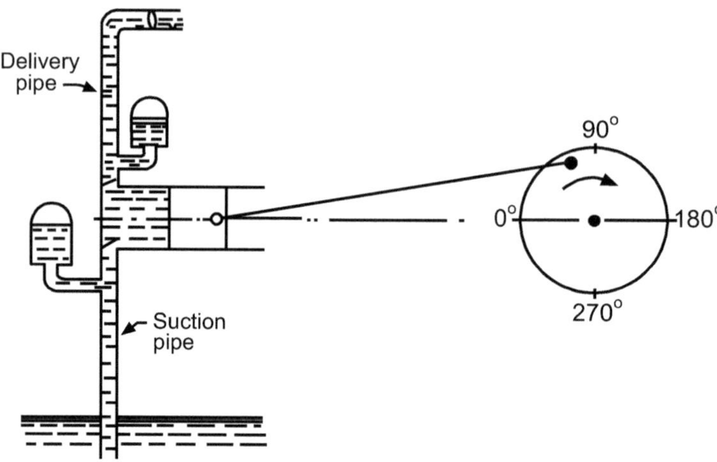

Fig. (5.33). Air vessels fitted to the suction and delivery pipes [4].

A static computation for obtaining the optimum size of the air vessel will consider useful volume of water stored in the tank - V_u In the hypothesis of a isothermal process - as presented in Fig. (**5.34**) - the relation for useful volume can be expressed as function of the pressures:

$$V_u = V_1 - V_2; \quad p_1 V_1 = p_2 V_2;$$

$$V_u = V_1\left(1 - \frac{p_1}{p_2}\right),$$

where: $V_1 = V_t$ – total volume in the tank; $p_1 = p_i$ - initial pressure of the gas; $p_2 = p_{max}$ – maximum pressure in the system.

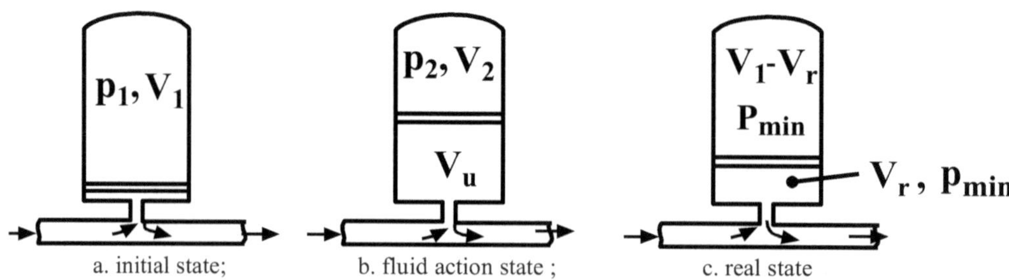

Fig. (5.34). The sketch for computing the air vessel.

Finally one can obtain the relation for the useful volume of the tank:

$$V_u = V_t\left(1 - \frac{p_i}{p_{max}}\right)$$

If the working regime of the air vessel is intense (high number of cycles) one considers a polytropic evolution of the gas and the relation becomes:

$$V_u = V_t\left(1 - \left(\frac{p_i}{p_{max}}\right)^{\frac{1}{n}}\right) \quad \textbf{(5.36)}$$

Note: In general the initial pressure in the tank is related to minimum pressure in the system therefore: $p_i = (0.8\ldots0, 9)\ p_{min}$ and for the real working conditions the air vessel is never completely discharged. For the minimum pressure in the system there is a residual liquid volume in the tank V_r. In considering the isothermal evolution between V_1; p_i to $(V_1\text{-}V_r)$; p_{min} then:

$$p_iV_1 = (V_1 - V_r)p_{min};\ V_r = V_1\left(1 - \frac{p_i}{p_{min}}\right) \text{ or } V_r = V_t\left(1 - \frac{p_i}{p_{min}}\right).$$

Finally one compute the real useful volume for the tank:

$$V_{u'} = V_u - V_r = V_t\left(\frac{p_i}{p_{min}} - \frac{p_i}{p_{max}}\right)$$

and:

$$V_{u'} = V_t\left(\left(\frac{p_i}{p_{min}}\right)^{\frac{1}{n}} - \left(\frac{p_i}{p_{max}}\right)^{\frac{1}{n}}\right) \quad \textbf{(5.37)}$$

if the evolution of the gas is polytropic.

The static computation for the air vessel could be valid if the length of pipes is short and small diameter. For the opposite cases one must take into account le oscillations of the system and therefore to avoid resonance phenomenon.

5.15. ROTARY POSITIVE DISPLACEMENT PUMPS

Rotary pumps are used primarily as a source of hydraulic fluid power and can be classified based on the types of rotating element – (Fig. **5.35**). It is self-priming and gives practically constant delivered capacity regardless of the pressure. Also rotary pumps are working as constant delivery pumps or variable delivery pumps.

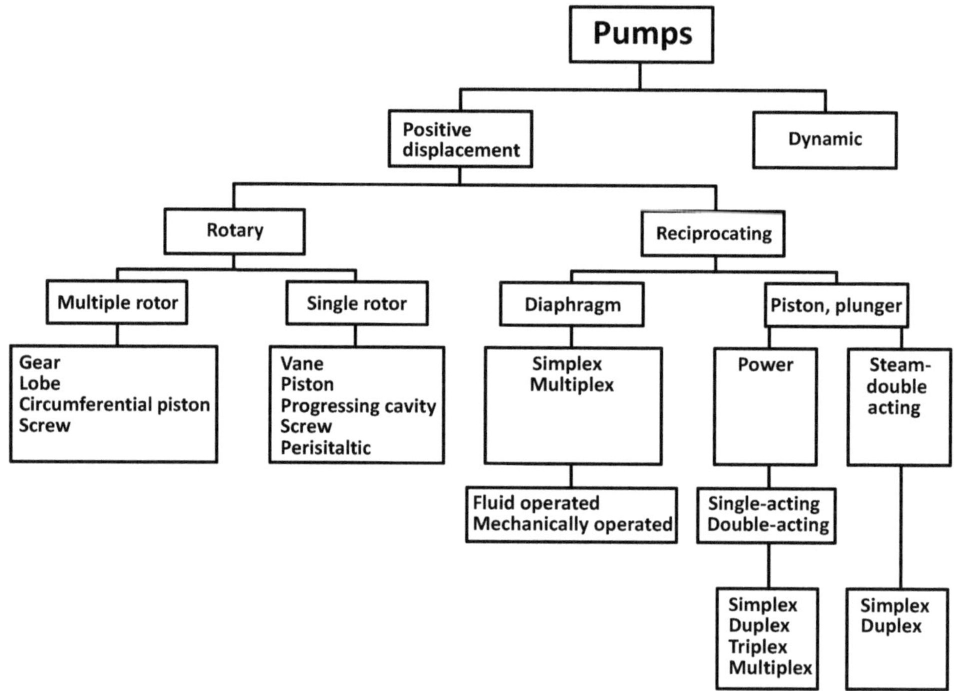

Fig. (5.35). Classification of rotary pumps [15].

Constant delivery pumps are classified with respect to the impeller element:

1- Gear pump - external type/internal type;

2- Screw pump

3- Vane pump.

4- Axial piston pump.

5- Radial piston pump.

5.15.1. Gear Pumps

A gear pump is a rotary pump consisting of two meshing gear wheels with contrarotation motion, where the fluid entrains on one side and discharges it on the other.

a) External type gear pumps - consists of two identical intermeshing gears (spur, Helical, Herring bore) – (Fig. **5.36**). It works with fine clearance inside a suitable shape casing. One gear is keyed to the driving shaft of a motor (driving gear) and the other revolves idly (driven gear). Oil entrains in the spaces between the gears, from suction port to the discharge port. Tighter internal clearances provides liquid passing through a pump for greater flow control.

They are available for:

-continuous pressure up to 250 bar;
-minimum speed 400 → 600 rpm;
-maximum speed 3000 → 6000 rpm;
-maximum delivery up to 10 l/s.

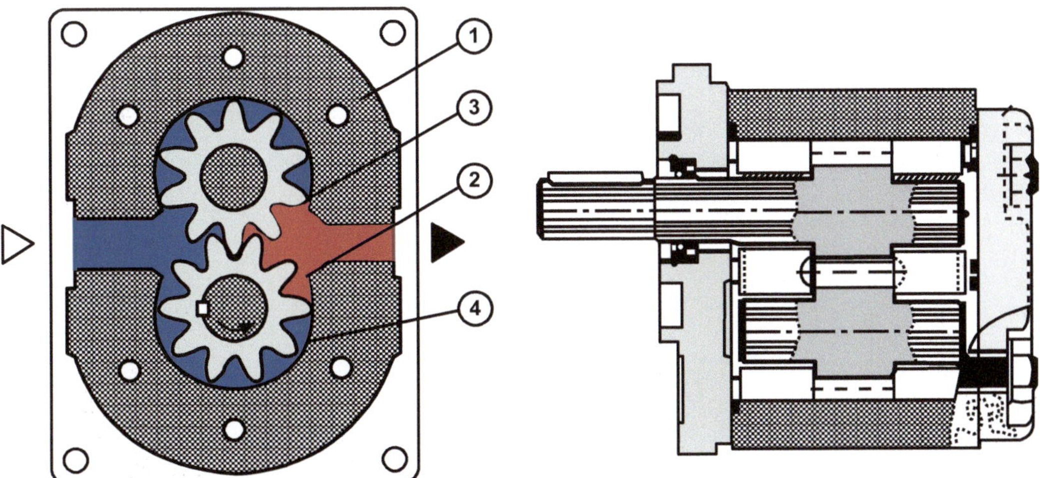

Fig. (5.36). External gear pump. 1- Pump body; 2- Idler gear; 3- Driven gear 4- Gear house [16].

Considering R_e and R_i as external and internal radius of the gear, the volume and flow rate of the pump are as follows:

$$V = \pi b(R_e^2 - R_i^2) \text{ and } Q = \pi bn(R_e^2 - R_{ie}^2)\eta_v \quad \textbf{(5.38)}$$

where: b – the width of the gear wheels; n – the rotation speed of the pump shaft; η_v = (0.8…8.85) volumetric efficiency.

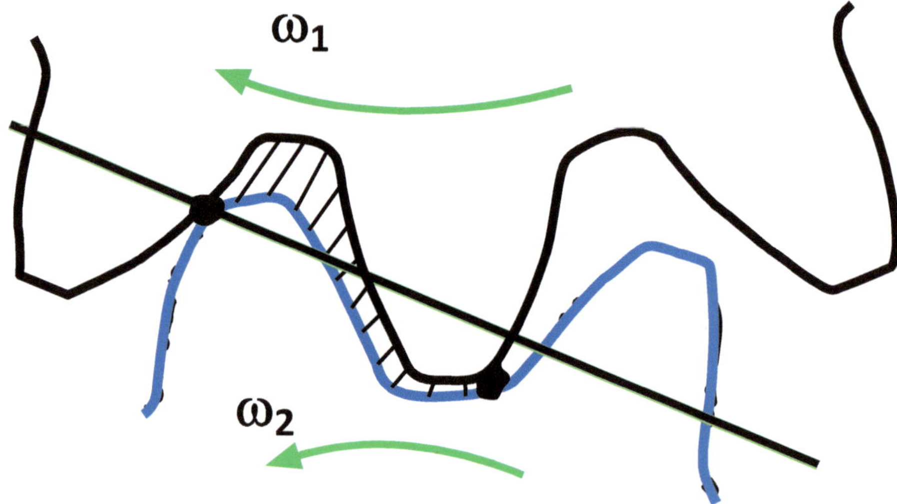

Fig. (5.37). The meshing gears contacting at two points.

The pressure corresponding to each active meshing gear is:

$$\delta p = \frac{p_2 - p_1}{z_a},$$

where: $(p_2\text{-}p_1)$ – static pressure for all active teeth of meshing gear; z_a active teeth of meshing gear. For a single couple of active teeth the force is:

$$F_1 = \delta p\, b\, (R_e - R_i).$$

Considering that F_1 is acting at the middle of the rectangle surface between the contacting points of the teeth – (Fig. **5.37**) *i.e.* $b\ (R_e\text{-}R_i)$ one obtains the torque for a single couple of active teeth:

$$M_1 = \frac{1}{2}\delta p\, b(R_e - R_i)(R_e + R_i) = \frac{1}{2}\delta p\, b(R_e^2 - R_i^2) \qquad \textbf{(5.39)}$$

Finally, the mechanical power for the pump with z_a active teeth is:

$$P = b\omega(p_2 - p_1)((R_e^2 - R_i^2) \qquad \textbf{(5.40)}$$

where: P = Mω.

The shape of the performance curves for a gear pump are presented in Fig. (**5.38**).

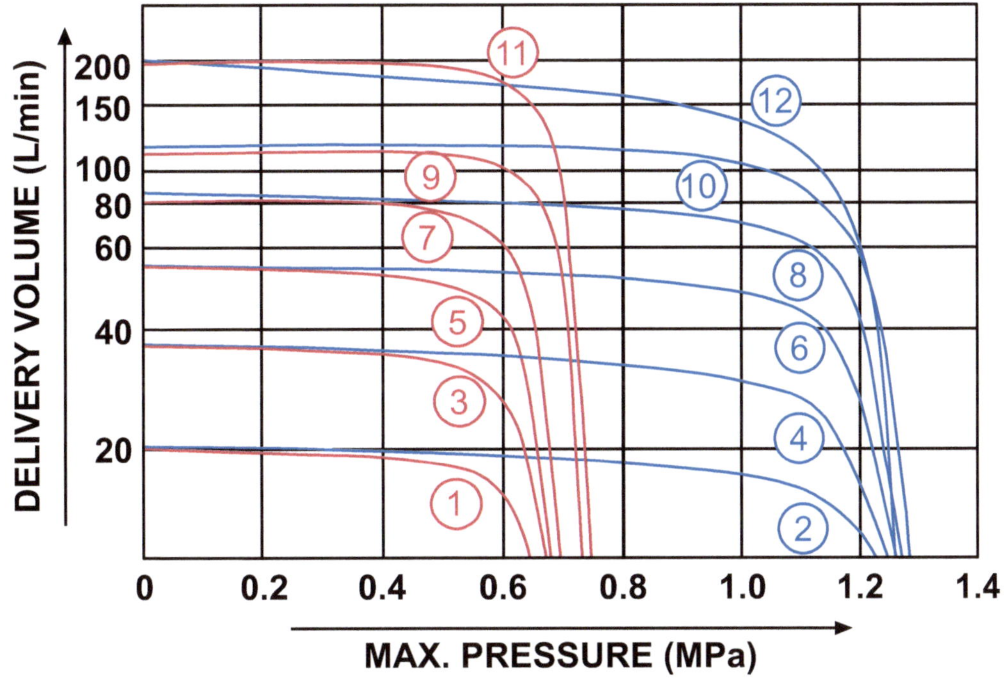

Fig. (5.38). Performance curves of gear pumps.

b) Internal type gear pumps

In this type the idler has an internal meshes with the driving gear – (Fig. **5.39**). The space between the outside diameter of idler is sealed by a crescent shape projection which from a part of the cover. The pump can be reversible if a provision is made for swinging the crescent through 180°.

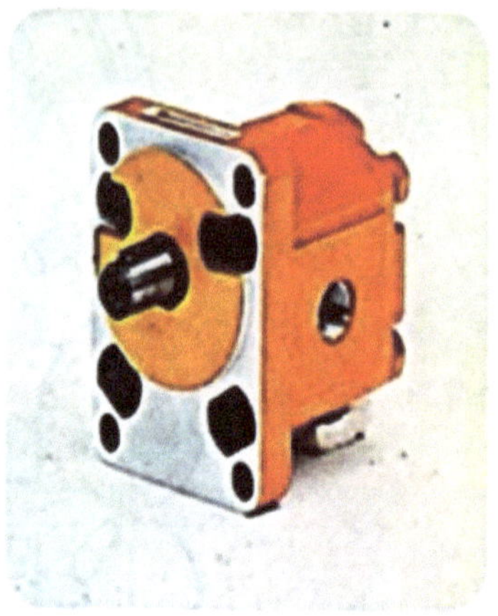

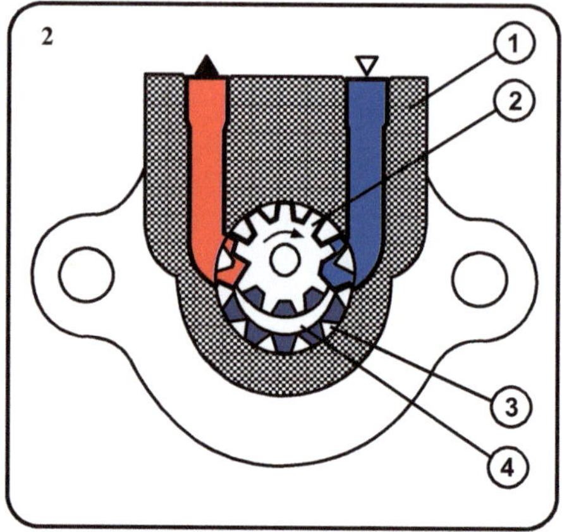

Fig. (5.39). Internal gear pump. 1- Pump body 2- Internal gear; 3- External gear 4- Crescent [16].

They are available for:

c) continuous pressure of 130 bar;
d) maximum speed 2000 →3600 rpm;
e) maximum deliveries of 4 l/s.

Advantages:

f) structure is simple;
g) light in weight;
h) size is small;
i) price is low;
j) ratio of trouble generation is small;
k) maintenance cost is small.

Disadvantages:

l) level of noise is high;
m) generates large pulsation.

5.15.2. Screw Pumps

It is an axial rotary pump consisting of a rotor directly connected to the source of power – (Fig. **5.40**). The rotor may be double helical (right and left hand) or single helical (one hand only). The fluid is carried from inlet section to the discharge along the rotor in pockets formed between teeth and the casing. There may be one, two or three screws. In a 2- screw pump, one is the rotor and the other is idler. In a 3-screw pump, there are two idlers on either side of the rotor.

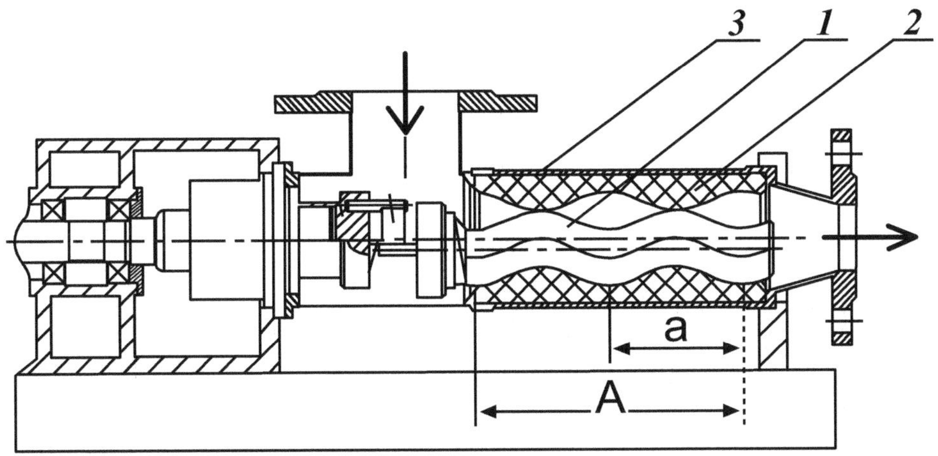

a

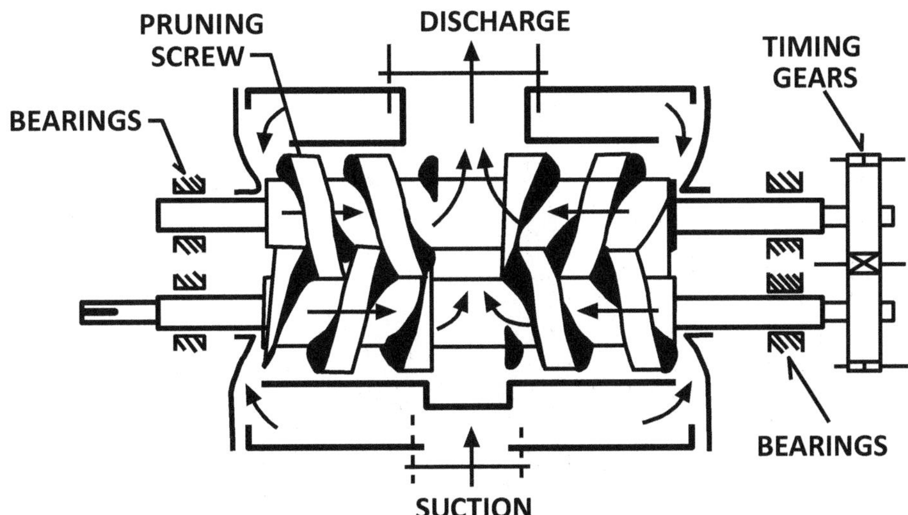

b

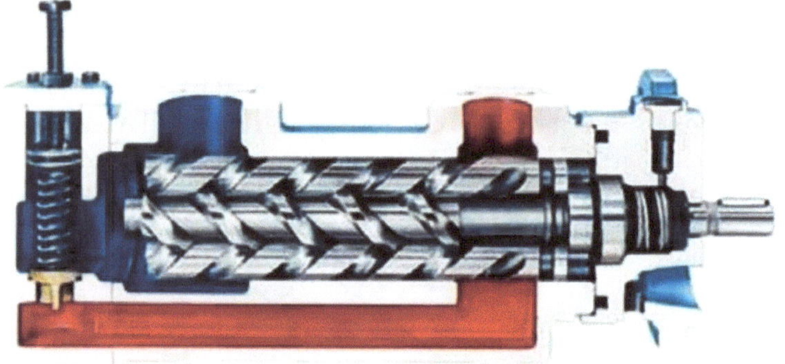

c

Fig. (5.40). Screw Pumps [1, 14]. (**a**). Single screw pump 1- Rotor; 2– Stator; 3-Casing; (**b**). Two- Screw Pump; (**c**). Three- Screw Pump.

For single screw pumps the flow rate can be computed by considering the geometry – (Fig. **5.41**) of the rotor and stator as follows:

$$Q = 2 \cdot 2e \cdot d \cdot s \cdot n \cdot \eta_v \qquad \textbf{(5.41)}$$

where: e – eccentricity; d – screw diameter; s – stator pitch; n = rotating speed; η_v = volumetric efficiency;

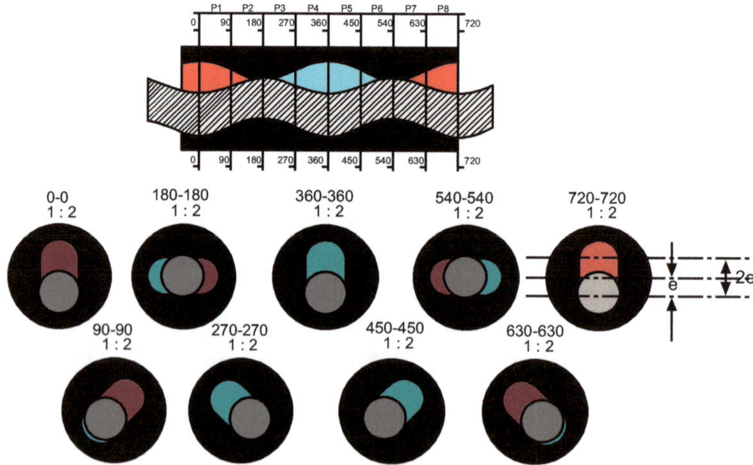

Fig. (5.41). The geometry of a single screw pump. (rotor pitch – 100 mm; stator pitch – 200 mm) [17].

There are available for.

n) continuous pressure up to 200 bar;
o) maximum speed 3000 → 4500 rpm;
p) maximum delivery up to 300 L/s.

Advantages

q) it is the quietest pump and free from vibration while running;
r) it is free from turbulence and pulsation.

Disadvantages

s) fluid of higher viscosity may require considerable derating the pump; the efficiency at 29 C° is 67% while at 50 C° is 54%.

The shape of the performance curve for a single screw pump are presented in Fig. (**5.42**).

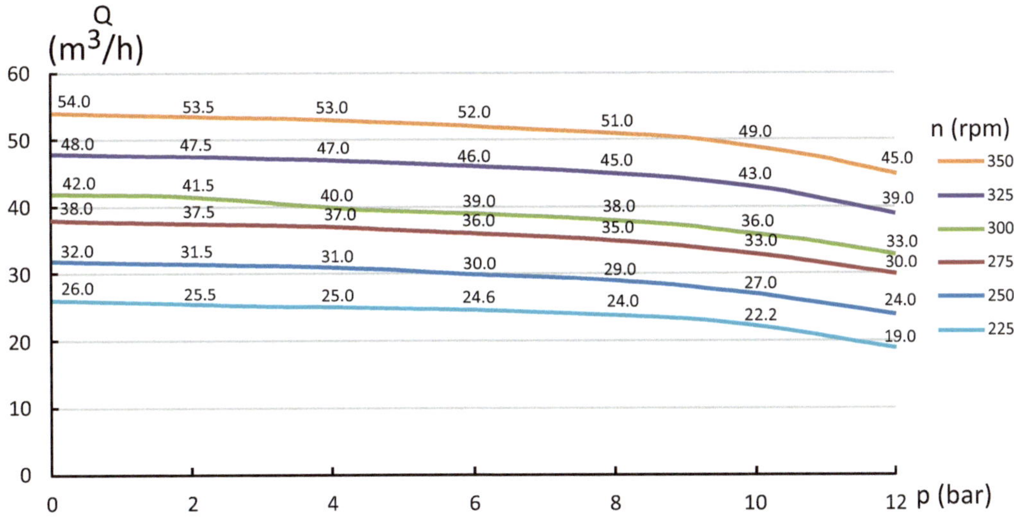

Fig. (5.42). Performance curves of screw pumps [17].

5.15.3. Vane Pumps

Rotary vane pumps – (Fig. **5.43**) consists of vanes mounted to a eccentrically installed rotor inside a cavity. The vanes (generally 4 → 8 vanes) creates variable cavities by moving under a spring force and following the casing wall. The vanes are free to slide radially with help of spring or hydraulic oil, thus obtaining the required seal between the suction and discharge connections.

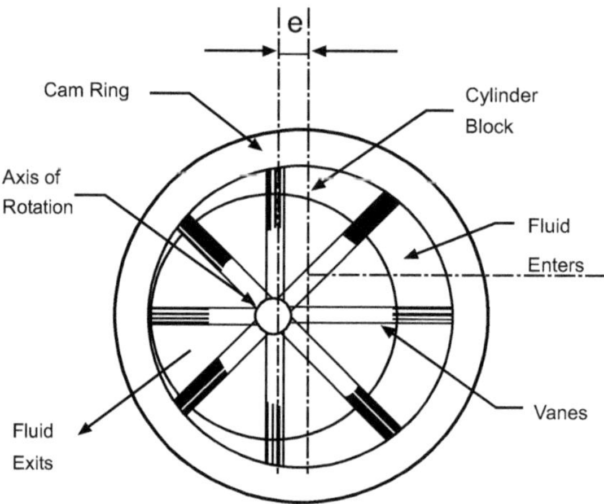

Fig. (5.43). Vane pump structure [13].

For vane pumps the flow rate can be computed by considering the geometry – (Fig. **5.44**) of the rotor and stator as follows:

$$Q = \left[\frac{\pi}{4}(D_e^2 - D_i^2) - zsl\right] bn\eta_v \quad \textbf{(5.42)}$$

where: D_e D_i – external and internal diameter; z – number of blades; s, l, b – dimensions of the vane; n – rotating speed; volumetric efficiency (0.94…0.98)%.

There are two types of vane pumps: fixed displacement unbalanced pump and fixed displacement balanced pump.

They are available for:

t) continuous pressure up to 160 bar;
u) speed 200 → 2500 rpm;
v) maximum delivery to 10 l/s.

Advantages:

w) pulsation of discharge pressure is small;
x) the shape and dimension are smaller in relation to other pumps;
y) repairing is easy;
z) the time required for attaining the maximum pressure is short.

Disadvantages

aa) working precision is high (high cost);
bb) the viscosity of oil to be employed is restricted.
cc) attention is required for the cleanness of oil.

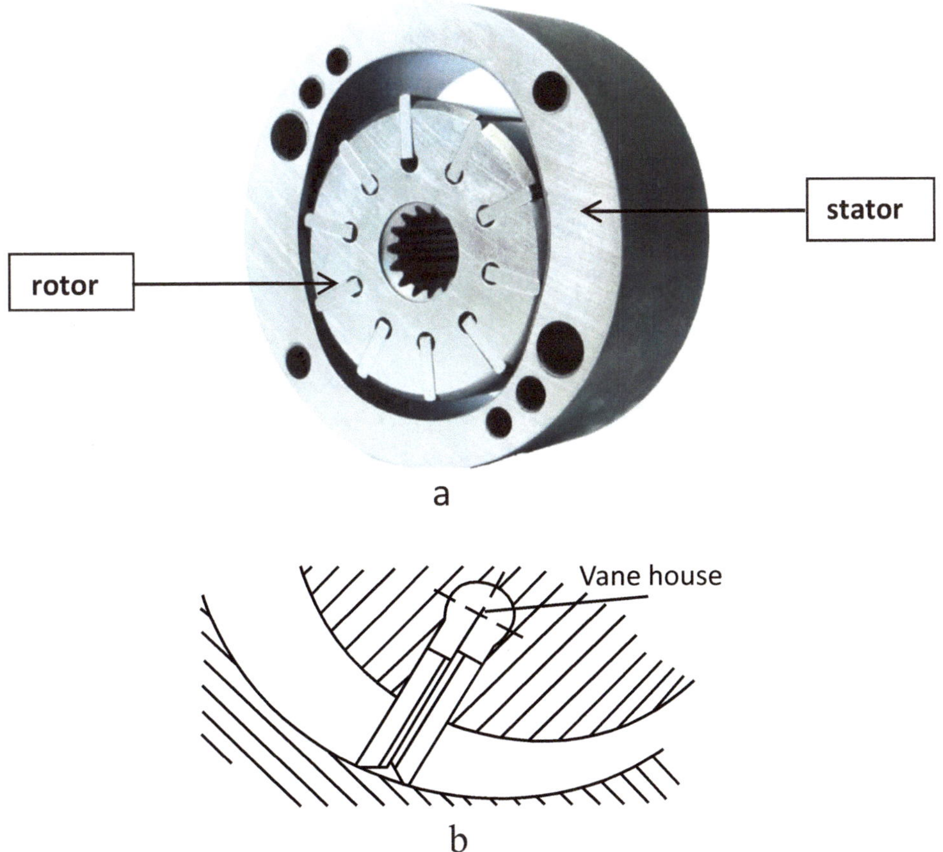

Fig. (5.44). Vane pump. (**a**) pump – rotor and stator [1]; (**b**) vane design - vane house.

The shape of the performance curves for a vane pump are presented in Fig. (**5.45**)

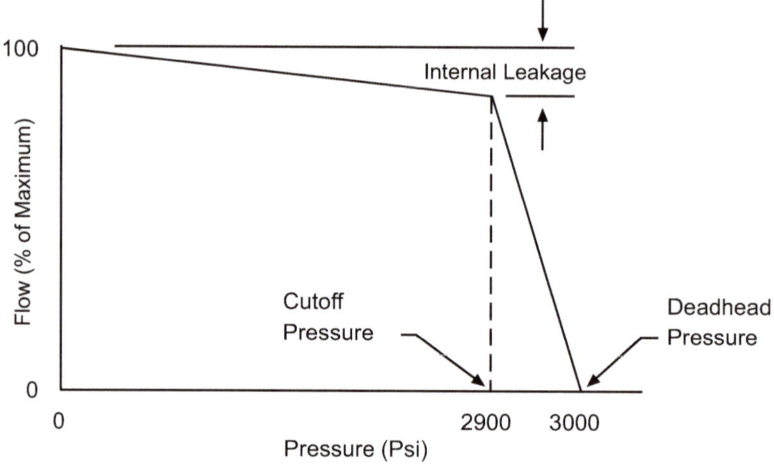

Fig. (5.45). Performance curve for a vane pump [13].

5.15.4. Axial Piston Pumps

The axial piston pumps type – (Fig. **5.46.a**) – are positive displacement pumps where the pistons are arranged parallel to the shaft of the pump rotor. As the driving shaft of the pump rotates, the cylinder barrel (which is inclined) provides variable moving chambers. When the cylinder block rotates, the piston shoes slides along the swash plate surface – (Fig. **5.45.b**). The piston stroke and the quantity of oil delivered are limited by the angle of the swash plate.

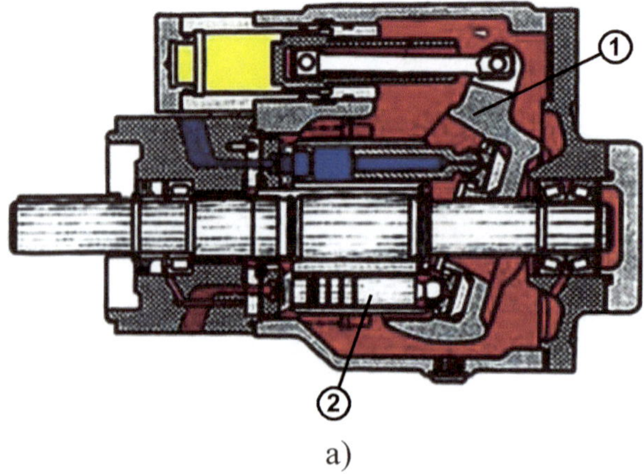

a)

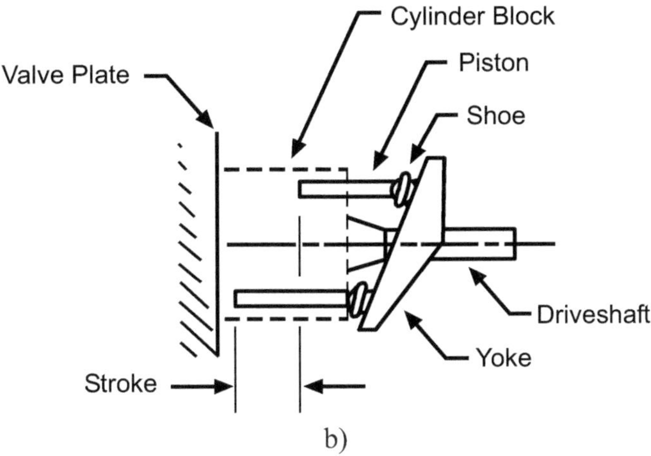

Fig. (5.46). Axial piston pump [1].

(**a**) axial piston pump: 1 -swash plate, 2 -piston; (**b**) sketch of an axial piston pump.

For axial piston pumps the flow rate can be computed by considering the geometry – (Fig. **5.47**) of the rotor and stator as follows:

$$Q = \frac{1}{4}\pi d^2 cNn = \frac{1}{4}\pi d^2 Dsin(\alpha)Nn\eta_v \quad \textbf{(5.43)}$$

where: D – diameter of the shoe; d – cylinder diameter; N – number of pistons; n – rotating speed; c – piston stroke; η_v – volumetric efficiency.

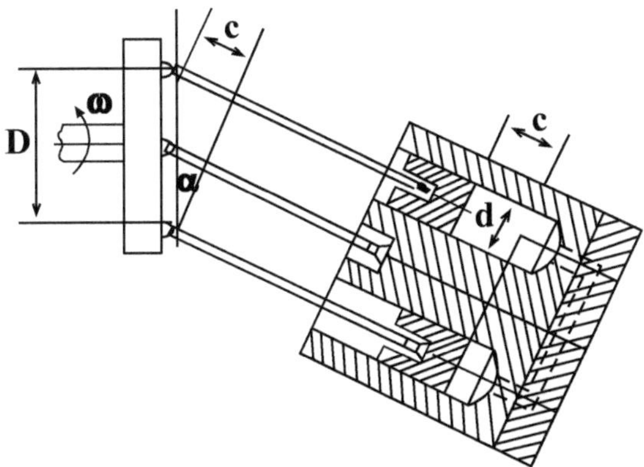

Fig. (5.47). Geometry of the axial piston pump [18].

The shape of the performance curves for an axial piston pump are presented in Fig. (**5.48**).

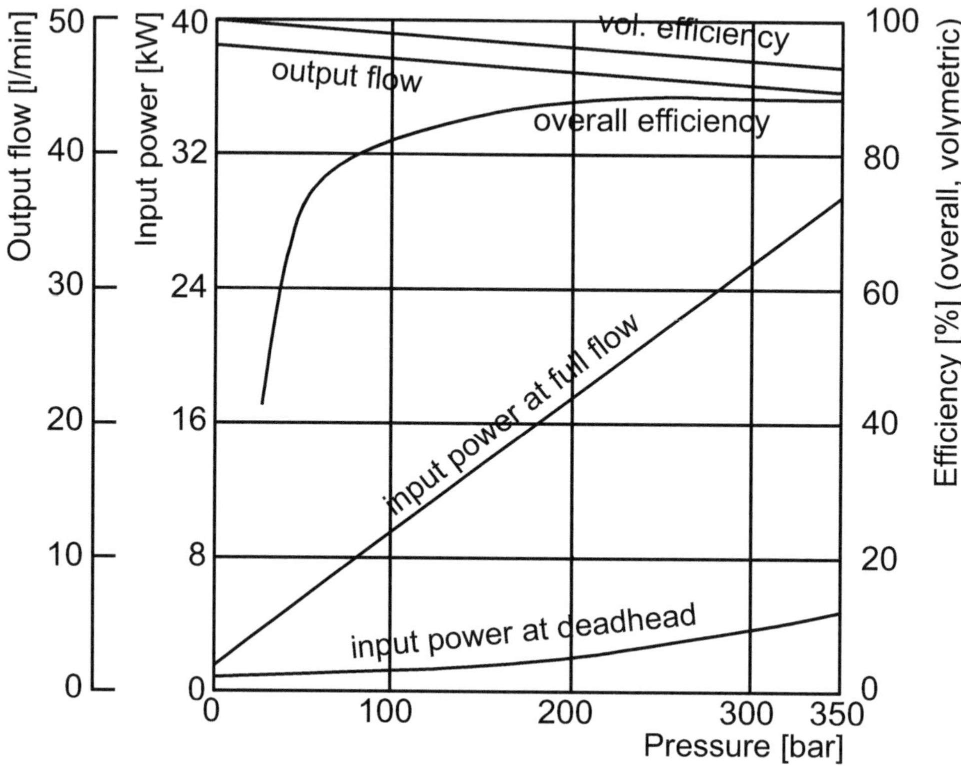

Fig. (5.48). Performance curves for an axial piston pump – PV 040 [19].

They are available for:

dd) continuous pressure 70 → 700 bar;
ee) rotating speed 900 → 1800 rpm;
ff) maximum delivery (1-500) l/s.

Advantages:

gg) the flow rate and the pressure can be controlled and depends on piston diameter;
hh) the pump can work in two direction which useful for mechanic tools.

5.15.5. Radial Piston Pumps

The radial piston pump consists of a number of cylinder displaced radially about an eccentric on the driving shaft – (Fig. **5.49**). The rotation of the cam shaft causes reciprocation of the piston. Number if piston (2- 24) which is a function of flow rate and pressure.

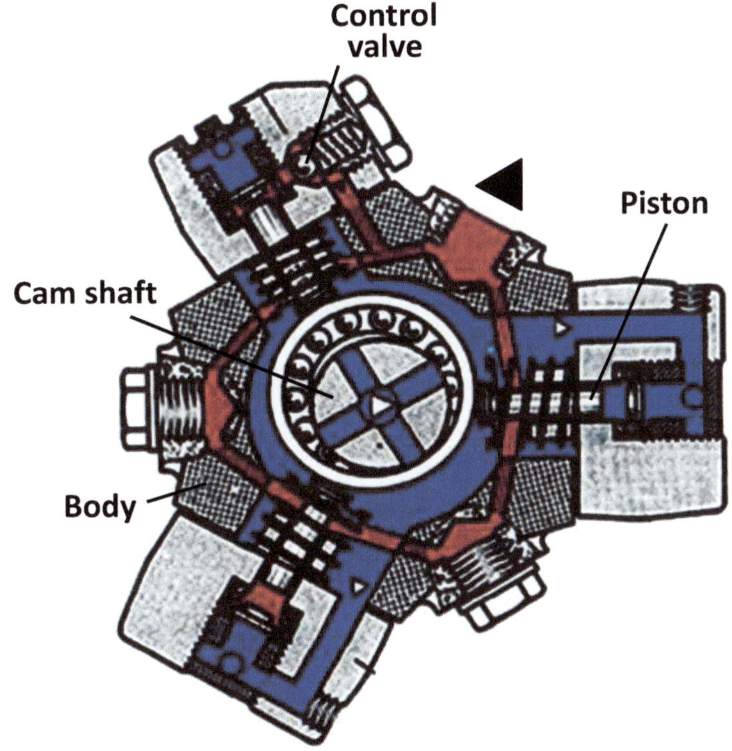

Fig. (5.49). Radial piston pump [16].

For radial piston pumps the flow rate can be computed by considering the geometry – (Fig. **5.50**) of the rotor and stator as follows:

$$Q = \frac{1}{4}\pi d^2 cnN\eta_v = \frac{1}{4}\pi d^2 2enN\eta_v \qquad \textbf{(5.44)}$$

where: d – cylinder diameter; N – number of pistons; n – rotating speed; c – piston stroke; e – eccentricity; η_v – volumetric efficiency.

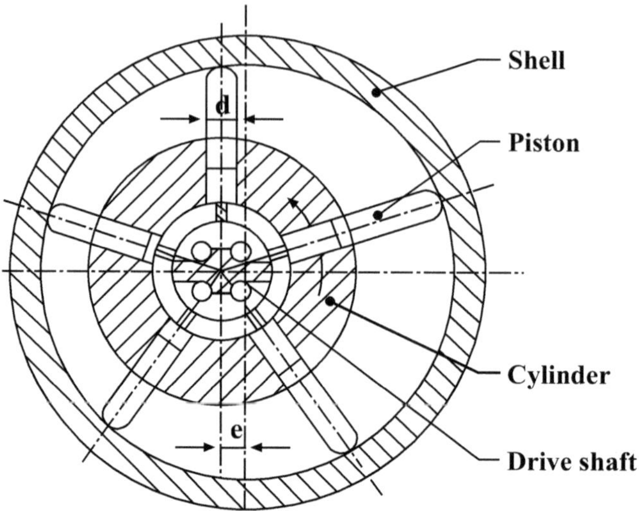

Fig. (5.50). Radial piston pump geometry [20].

The shape of the performance curves for a radial piston pump are presented in Fig. (**5.51**).

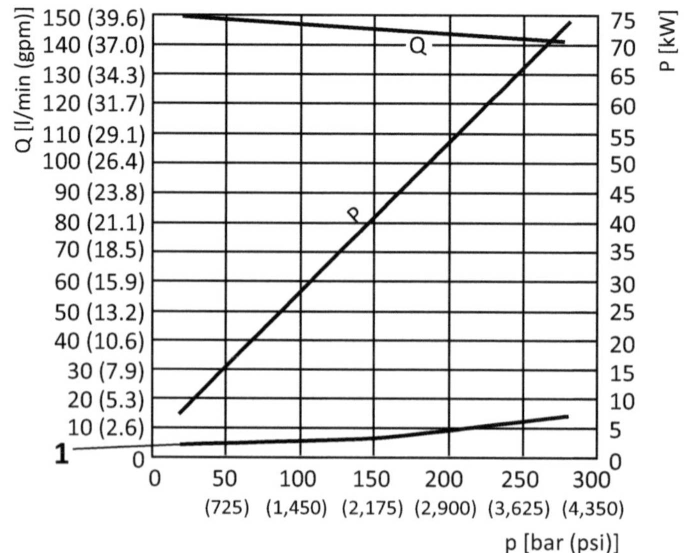

Fig. (5.51). Performance curves for a radial piston pump – V = 100 cm^3/rev; (1- P at zero stroke) [20].

They are available for:

ii) continuous pressure up to 650 bar;
jj) rotating speed up to 3200 rpm;
kk) maximum delivery up to 200 l/min.

In general, all the pumps are used as a variable delivery pump by controlling the flow rate and pressure *e.g.* the swash plate could be radial type at a variable angle, the rotor of radial type at a variable speed *etc.*

5.15.6. General Pumping Formulas

The computation of pumping in hydraulic installations are based on general pumping formulas taking into account the different conditions of the application.

a- the ratio of the output from a pump to the input, describing overall efficiency

$$\eta_{overall} = \frac{\text{Pump output(Fluid power)}}{\text{Pump input(Brake power)}}$$

$$\eta_o = \frac{P \times Q}{T \times \omega} \quad \textbf{(5.45)}$$

Ex: A pump operatives at 174 bar and 22× 10^{-5} m^3/s. the input power 4 kW compute the overall efficiency and the input Torque to the pump when it run at 1725 rpm.

Solution :-

$$\eta_o = \frac{P \times Q}{T \times \omega} = \frac{174 \times 10^5 \times 22 \times 10^{-5}}{4 \times 1000} = 96\%$$

$$also\ \eta_o = \frac{P \times Q}{T \times \omega} \quad \therefore\ T = \frac{174 \times 10^5 \times 22 \times 10^{-5}}{0.96 \times 1725 \times \frac{1}{60} \times 2}$$

$$T = \frac{138.7}{2\pi}\ N.m = 22.1\ N.m$$

b- Theoretical Pump Torque is expressed in term of pump displacement and specified operating pressure:

$$T_{th} = \frac{Displacement(V_p) \times Pressure(P)}{2\pi} \quad (5.46)$$

$$T_{th} = \frac{V_p \times P}{2\pi}$$

V_p = Volume displaced per revolute = CC/rev or l/rev

$$Torque\ efficiency = \frac{Tactual}{T_{th}} \times 100$$

Which is the same as mechanical efficiency.

c- Leakage or pump cavity flow in positive displacement hydraulic pumps is computed in liter/min

$$Volumetric\ efficiency\ \eta_Q = \frac{Pump\ volume\ output}{Pump\ displacement \times Speed}$$

$$\eta_Q = \frac{Q_o}{V_p.N} \times 100 \quad (5.47)$$

Then

$$Pump\ overall\ efficiency = Pump\ volumetric\ eff. \times Pump\ Mechanical\ eff.$$

$$\eta_{overall} = \eta_Q . \eta_T$$

d- Frictional power losses are computed as a portion of the brake power at pump drive shaft.

$$Pump\ Frictional\ Power = Pump\ Brake\ Power[1 - Power\ Mech.efficiency]$$

$$FP = BP[1 - \eta_T]$$

For gear pump

Let

A- The area enclosed between Two adjacent teeth and casing.

l- The axial length of teeth.

n- Number of teeth in each pinion.

N- The speed in rpm.

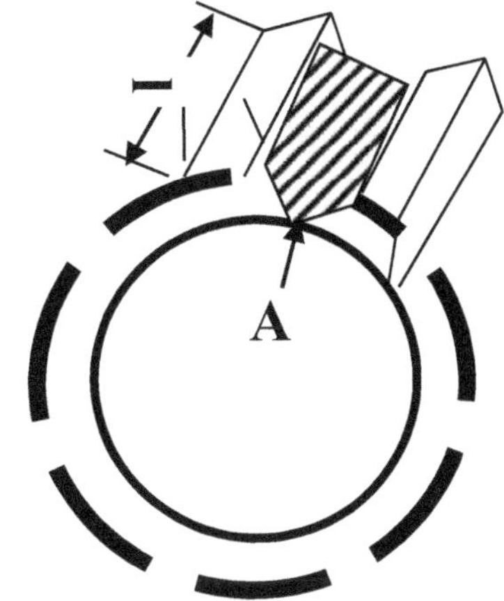

$$\therefore \text{The volume of liquid pumped in one revolution} = 2Al.n$$

$$Or\ V_p = 2Al.n \text{ for external type}$$

$$\therefore \text{Total or ideal discharge} = \frac{2AlN}{60}$$

$$If\ \eta_Q \text{ be the Volumetric efficiency}$$

$$\therefore\ Q_{act} = \frac{2AlN}{60}.\eta_Q$$

$$or\ Q_{act} = V_p.N.\eta_Q$$

$$Note: \text{for internal type } V_p = Al.n$$

$$n = \text{Number of teeth(total)}$$

The theoretical displacement of gear pump is

$$V_p = 2(A - A_m)l.n \tag{5.48}$$

A_m = The area between the meshing teeth

It is difficult to determine A_m the following empirical relation could be used.

$$V_p = 0.95\ \pi . C. (D - C)l \quad \textbf{(5.49)}$$

$C -$ *The center to center distance between gear axis's.*

$D -$ *Outside diameter of gears.*

$n -$ *Number of teeth related to the ratio* $\frac{D}{C}$

n	7	10	13	18
D/C	1.28	1.21	1.13	1.12

For Piston Pump

A- The cross- section area of the piston or cylinder

S- Stroke

$$\therefore Cylinder\ Volume = A.S$$

$$\therefore V_p = A.S$$

$$or\ Q_{act} = V_p.N.\eta_Q$$

$N -$ *Speed rpm.*

SOLVED PROBLEMS

Q.1- Compute the volumetric efficiency of a pump that has a positive displacement of 61.5 cm^3 and delivers 3.28 l/s of fluid while operating at 3300 rpm.

Solution:-

$$\eta_Q = \frac{Q}{V_p \times N} = \frac{3.28}{61.5 \times 10^{-3} \times \frac{3300}{60}}$$

$$= 0.97\ or\ 97\%$$

Q.2- A hydraulic pump is observed to have an overall efficiency of 87% while consuming 7.457 kW Brake power measured at the pump drive shaft. Compute the mechanical efficiency and the frictional power loss from the system when the volumetric efficiency 94%.

Solution:-

$$\eta_T = \frac{\eta_{overall}}{\eta_Q} = \frac{0.87}{0.94} = 0.9255$$

$$FP = 7.457[1 - 0.9255] = 0.555\ kW$$

Q.3- A) Determine the output torque of a hydraulic motor with a displacement of 60 cm^3/rev when the pressure drop across the motor is 100 bar (assume 100% torque efficiency). B) Determine the SIKW power delivered to the load of the motor is 600 rev/min, and the overall efficiency of the motor is 100%.

Solution:-

A)

$$Torque\ (N.m) = \frac{P\left(\frac{N}{m^2}\right).V_p\left(\frac{m^3}{rev}\right)}{2\pi\left(\frac{rev}{rad}\right)}$$

$$= \frac{100 \times 10^5 \times 60}{2\pi \times 10^6}$$

$$= 95.5\ N.m$$

B)

$$Power = T.W \qquad W = \frac{2\pi N}{60}$$

$$= 95.5 \times \frac{2\pi N}{60}$$

$$= 6\ kW$$

Q.4- What are the input power and the overall efficiency of internal gear type pump when the speed 900 rpm, torque 40 N.m, the flow rate 10 l/min and the pressure 200 bar?

Solution:-

$$(Mech\ rotary) or\ input\ power = T.W$$

$$P_{in} = 40 \times \frac{2\pi N}{60}$$

$$= 40 \times \frac{2\pi \times 900}{60} = 3.77\ kW$$

$$\eta_{overall} = \frac{P_{out}}{P_{in}} = \frac{P \times Q}{3.77}$$

$$= \frac{200 \times 10^5 \times 10 \times \left(\frac{1}{1000 \times 60}\right)}{3.77}$$

$$= \frac{3.33}{3.77} \times 100 = 88\%$$

Q.5- Calculate the following efficiency for an assumed pumping system of internal pump.

a) Electric drive motor efficiency.

b) Hydraulic pump efficiency.

c) System overall efficiency.

Use the following input and outputs.

1- Motor input power = 3.73 kW

2- Motor output power = 4.5 hp × 0.745 = 3.357 kW

3- Pump input power = 4.5 hp× 0.746 =3.357 kW

4- Pump output flow rate = 441.6 cm^3/s

5- Pressure output = 68 bar

Solution:-

a)

$$Electric\ motor\ efficiency = \frac{P_{out}}{P_{in}}$$

$$= \frac{3.357}{3.73} = 90\%$$

b)

$$Pump\ overall\ efficiency = \frac{P_{out}}{P_{in}}$$

$$= \frac{Pressure \times Flow\ rate}{Pump\ input\ power(Electrical\ motor\ output)}$$

$$= \frac{68 \times 10^5 \times 441.6 \times 10^{-6}}{3.357 \times 1000} = 89.5\%$$

c)

$$System\ overall\ efficiency = \frac{Pump\ output}{Electrical\ motor\ input}$$

$$Pump\ output = P.Q = 68 \times 10^5 \times 441.6 \times 10^{-6} \times \frac{1}{1000}$$

$$= 3\ kW$$

$$\therefore \ System\ overall\ efficiency\ \eta_o = \frac{3}{3.73} \times 100$$

$$= 80.4\%$$

$$or\ \eta_{overall} = \eta_{motor} \times \eta_{pump}$$

$$= 0.9 \times 0.895$$

$$= 0.8055\ \ or\ \ 80.55\%$$

Q.6- What will be the hydraulic power delivered by external gear pump if it supplies 315.5 cm^3/s and P= 117 bar. If the overall efficiency of the pump and the motor is 78%. What mechanical power output of motor to drive it.

Solution:-

$$Hydraulic\ power\ delivered\ by\ the\ pump\ = PQ$$

$$= 117 \times 10^5 \times 315.5 \times 10^{-6}$$

$$= 3685.5\ Watt$$

$$\eta_{overall} = \frac{P_{output}}{P_{input}} = \frac{P_{out}}{3685.5} = 0.78$$

$$P_{out}\ of\ the\ motor = 2874.7\ Watt$$

Q.7- What will be the overall efficiency of an internal type gear pump for N= 1200 rpm, T = 100 N.m, Q = 28 l/min and P = 240 bars.

Solution:-

$$\eta_o = \frac{P_{out}}{P_{in}} = \frac{P\,Q}{T\,\omega}$$

$$= \frac{240 \times 10^5 \times 28 \times \frac{10^{-3}}{60}}{100 \times \frac{2\pi N}{60}}$$

$$= \frac{240 \times 10^5 \times 28 \times 10^{-3} \times 60}{100 \times 2\pi \times 1200 \times 60}$$

$$\eta_o = 89\%$$

Q.8- A hydraulic motor produces 7.5 kW at a speed of 150 rod/s with a pressure loss of 10 bar from the pump to motor. Determine the input power to the pump, pump flow rate, output torque when the efficiency of the pump and the motor are.

Efficiency	Pump	Motor
η_V	0.97	0.9
η_T	0.95	0.85

System pressure 90 bar

Leakage in system 6 l/min

Solution:-

$$P_{out} = T.W$$

$$The\ motor\ torque\ output = \frac{Power\ out}{Speed}$$

$$or\ = \frac{7.5 \times 10^3}{150} = 50\ N.m$$

$$Theoretical\ output\ Torque = \frac{actual\ Torque}{\eta_T}$$

$$= \frac{50}{0.85} = 59\ N.m$$

$$Pressure\ available\ at\ the\ motor = Pressure\ at\ the\ pump - losse$$

$$= 90 - 10 = 80\ bar$$

$$\therefore Theoretical\ flow\ rate\ at\ the\ motor = \frac{T_{th} \times W}{Pressure}$$

$$= \frac{59 \times 150}{80 \times 10^5} = 1.1 \times 10^{-3}\ m^3/s$$

$$\therefore \text{Actual flow rate at the motor} = \frac{1.1 \times 10^{-3}}{\eta_V}$$

$$= \frac{1.1 \times 10^{-3}}{0.9} = 1.23 \times 10^{-3}\ m^3/s$$

$$Flow\ out\ of\ the\ pump = Actual\ flow\ at\ the\ motor + System\ leakage$$

$$= 1.23 \times 10^{-3} + \frac{6 \times 10^{-3}}{60} = 1.33 \times 10^{-3}\ m^3/s$$

$$P_{out} \text{of the pump} = \text{PQ}$$

$$= 90 \times 10^5 \times 1.33 \times 10^{-3}$$

$$= 12 \text{ kW}$$

$$P_{in} \text{of the pump} = \frac{12}{\eta_o} = \frac{12}{0.47 \times 0.95} = 13 \text{ kW}$$

$$\text{Theoratical flow of the pump} = \frac{1.33 \times 10^{-3}}{0.97}$$

$$= 1.38 \text{ l/s}$$

$$or\ The\ pump\ input\ (neglecting\ the\ losses) = \frac{motor\ output}{\eta_{Vm}\eta_{Tm}\eta_{Vp}\eta_{Tp}}$$

$$= \frac{7.5 \times 10^3}{0.9 \times 0.85 \times 0.97 \times 0.95} = 10.64\ kW$$

Q.9- A hydraulic pump is operating at 10.3 MPa pressure and 37.85 l/min flow rate. The pump is driven by an electric motor at 1725 rpm. The electric motor produce 3.95 kN-m of torque. Find the overall efficiency of the pump.

Solution:-

$$\eta_{overall} = \frac{P_{out}}{P_{in}}$$

$$P_{out} = \gamma QH = P \times Q$$

$$= 10.3 \times 10^6 \times 37.85 \times \frac{1}{1000 \times 60}$$

$$P_{in} = T.W = 3.95 \times 1000 \times \frac{2\pi N}{60}$$

$$= \frac{3950 \times 2\pi \times 1725}{60}$$

$$\eta_o = \frac{10.3 \times 10^6 \times 37.85 \times \frac{1}{60000}}{3950 \times 2\pi \times 1725 \times \frac{1}{60}}$$

$$= 0.91 \quad or \quad 91\%$$

REFERENCES

[1] J.M. Hassan, *Power Generation.* University of Technology, Mech. Eng. Dept., 2018.

[2] Q.H. Nagpurwala, *Hydraulic Turbines.* MS Ramaiah School of Advanced Studies: India, 2015.

[3] M. Manno, *Hydraulic turbines and hydroelectric power plants. Energy Systems Course, Lecture Notes.* Department of Industrial Engineering, University of Rome, 2013.

[4] J. Lal, *Hydraulic Machines: Including Fluidics,* 6th ed. Metropolitan Book Co. (P) Ltd: New Delhi, 2016.

[5] G.I. Krivchenko, *Hydraulic Machines Turbines and Pumps,* MIR publishers Moscow: Kochin, 1986.

[6] E.H. Rachael Haas, *Michael Hiebert, Francis Turbines Fundamentals and Everything Else You Didn't Know That You Wanted to Know.* Colorado State University, 2014.

[7] Jacobsen Christian Brix, *The Centrifugal Pump* Grandiose Search and Technology.

[8] Pumps.org, *Hydraulic Institute,* 2019. Available at: http://pumps.org (Accessed: 9th April, 2018).

[9] *Hydraulic Training Course.* Technical & Vocational Training Corporation: Saudi Arabia, 2003.

[10] Ge.com, "GE Power | General Electric", Available at: https://www.ge.com/power (Accessed: 11th May, 2018).

[11] J.M. Chapallaz, P. Eichenberger, and G. Fischer, *Manual on Pumps Used as Turbines,* Vieweg: Braunschweig, Germany, 1992.

[12] "Engineering notes India", *Essays, Research Papers and Articles on Engineering Notes India.* 2019. Available at: http://www.engineeringenotes.com (Accessed: 20th July, 2018).

[13] Pumpschool.com, Available at: http://www.pumpschool.com (Accessed: 7th May, 2018).

[14] H. Exner, R. Freitag, I.H. Geis, R. Lang, J. Oppolzer, P. Schwab, and E. Sumpf., "Instruction and information on the basic principles and components of fluid technology", *Mannesmann Rexorth,* 1991.

[15] Confind, *Confind.ro,* 2019. Available at: http://www.confind.ro (Accessed: 11th May, 2018).

[16] "Hydraulic Schematic Troubleshooting", *Hydraulicstatic.com,* 2019. Available at: http://www.hydraulicstatic.com (Accessed: 5th May, 2018).

[17] "DirectIndustry - El salón online de la industria: sensores, automatismos, motores, bombas, manipulación, embalajes", *Directindustry.es,* 2019. Available at: https://www.directindustry.es (Accessed: 11th August, 2018).

[18] "Hydraulic Pump", *Hydraulic-pump.info,* 2019. Available at: http://www.hydraulic-pump.info (Accessed: 11th August, 2018).

[19] Parker Hannifin, Available at: https://www.parker.com/portal/site/PARKER/menuitem.223a4a3cce02eb6315731910237ad1ca/?vgnextoid=c302e75fd272f210VgnVCM10000048021dacRCRD&vgnextfmt=EN (Accessed: 12th January, 2018).

[20] M. Inc, "Moog, Inc. - Precision motion control products, systems, servovalves, actuators", *Moog.com,* 2019. Available at: https://www.moog.com (Accessed: 12th February, 2018).

Jafar Mehdi Hassan, Salman Hussien Omran, Laith Jaafer Habeeb, Alamaslamani Ammar Fadhil Shnawa & Adrian Ciocănea

SUBJECT INDEX

A

Accelerating head 182
supplying 182
Adjustable-blade runner 77
Air vessels 174, 175, 182, 183, 184
Angle 69, 84, 130, 163, 200
cone 84
exit 163
optimum 130
theoretical inlet 69
variable 200
Application 4, 129
of momentum equation 4
pumping 129
Appling Bernoulli's equation 58
Atmospheric 34, 49, 86, 139, 141
air 34
pressure 49, 86, 139, 141
Axial 148, 189
rotary pump 189
Axial flow 115, 127, 161
pumps 127, 161
Axial flow turbines 75, 118
law-head 75
Axial piston pump 185, 195, 196, 197
type 195

B

Balancing piston 34
Barometric head 112
Bearings 75, 114
blade pin 75
Bernoulli's equation 1, 11
Blade 75, 78, 104
angle 75, 104
flange 75
levers 78
pivot 75
Bolt head 34
Breadth ratio 63, 64
Brealth of runner 94
Bronze sleeve 34
Bubble cavitations 108, 109

C

Casing 33, 34, 50, 86, 130, 132, 190, 193, 202
hood 34
of centrifugal pumps 130
scroll 50
spiral 50, 130
volute 130
wall 193
Cavitation 107, 108, 139, 143, 144, 182
avoiding 182
conditions 107, 108
inception 139, 144
phenomenon 144
risk 143, 144
Cavities 107, 144, 174, 193
gas-filled 107
Centrifugal pumps 124, 126, 127, 128, 130, 131, 134, 136, 139, 147, 149, 150, 158, 159, 160, 163, 175
medium lift 126
multistage 127
ordinary 128
stage 127
Channel flow 4
Characteristics 32, 84, 177
power-generating 32, 84
of reciprocating pumps 177
Classification 103, 120, 122, 124, 125, 185
hydraulic turbine 103, 120, 122
Closed 128, 150, 152
loop 150, 152
type impeller 128
Coefficient 24, 37, 38, 40, 59, 69, 87, 146, 177, 178, 180
blade thickens 69
dimensionless power 146
of speed ratio 24, 37

Jafar Mehdi Hassan, Salman Hussien Omran, Laith Jaafer Habeeb,
Alamaslamani Ammar Fadhil Shnawa & Adrian Ciocănea

speed 87
Conditions 107, 114, 147
hydrodynamic 147
operating 107, 114
Connecting rod mechanism 177, 178
Constant 11, 41, 64, 65, 95, 96, 139, 146, 169, 185
impeller diameter 146
radial flow velocity Vf 139
Construction of reaction turbine 50

D

Damages 107, 108, 109
edge cavitation 109
Delivery 104, 130, 134, 135, 142, 163, 164, 165, 175, 176, 177, 179, 182, 183
gauges 134
head 134
pipes 130, 135, 142, 163, 164, 176, 182, 183
strokes 177, 179
unit power 104
Design 24, 25, 32, 34, 75, 105, 107, 124, 140, 148
hydraulic 107
pump inlet 140
Diameter 18, 19, 23, 24, 25, 30, 37, 38, 40, 60, 64, 65, 67, 71, 73, 79, 90, 91, 92, 99, 101, 158, 163, 169, 172, 178, 180, 184, 191, 193, 196, 198
bottom 60
circle 30
cylinder 196, 198
external 73, 99, 101
inner 64, 67, 91
internal 172, 193
least 23, 24, 25, 37
nominal 158
of Pelton wheel 24
outer 65, 92
pipe 178
pipeline 40
screw 191
small 184
Dimensional 103, 131
and model analysis 103
fluid flow 131
Dimensionless coefficients 146
Discharge 21, 23, 33, 41, 49, 57, 64, 65, 79, 90, 92, 97, 158, 171, 175, 177, 179, 180, 181, 182, 186, 190, 193, 194
coefficient of 180, 181
connections 193
flow, effective 177
nozzle 158
pressure 194
radial 90, 92
reciprocating pump 182
submerged 57
systems 21, 33, 49, 79
Discharging cooling water 34
Displacement 75, 193, 202, 204
fixed 193
theoretical 202
Draft tube 49, 50, 51, 57, 58, 59, 60, 61, 84, 85, 98, 107, 108
cones 107
efficiency 59
exit diameter 84
outlet 57, 98
straight-type conical 84
theory 108
Drainage channels for spray water 34
Dynamic 108, 136
head 108
sealing 136
Dynamic force 1, 5, 52
resultant 52

E

Effective dynamic suction head 108
Effect 146
of diameter variation 146
of speed variation 146
Efficiencies of centrifugal pumps 136
Efficiency of reaction turbines 53

Elblow 84
  tube 84
  type draft tubes 84
Electric drive motor efficiency 205
Empirical formula 108
Emptying pipe 34
Energy 44, 49, 86, 88, 123, 124, 126, 151
  consumption 151
  mechanical 124
  systems 123
Equation 3, 63, 82, 130, 132, 133
  empirical 82
  fundamental 132
Evolution 184
  isothermal 184
  polytropic 184

F

Flow 1, 4, 5, 6, 7, 49, 52, 57, 92, 103, 127, 162, 163, 186
  area 32
  characteristics 4
  conditions 162
  constant 92
  control 186
  direction 4, 5, 6, 7, 49, 57, 127
  passages 1, 4, 52
  phenomena 103
  radial 163
Flow rate 26, 146, 182, 206
  coefficient 146
  oscillations 182
  pump output 206
  supplied water 26
Fluid 1, 2, 3, 4, 5, 11, 12, 103, 118, 124, 126, 132, 137, 139, 140, 144, 183, 186, 190, 192
  acceleration 12
  action state 183
  control volume 2
  dynamic force 1
  entrains 186
  mechanics 103
  motion 1, 11
  pumped 140
  pressure 2, 11
  velocity 144
Fluid power 185, 200
  hydraulic 185
Force 1, 2 3, 4, 5, 8, 11, 12, 18, 19, 21 30, 31, 75, 125, 136, 193
  and power of centrifugal pumps 136
  centrifugal 11, 12, 125
  fluid dynamics 1
  frictional 75
  gravitational 12
  gravity 12
  impulsive water 21
  pressure difference 12
  real applied external 2
  spring 193
Forged steel 128
Francis 49, 50, 64, 107, 118, 119
  runner 50
Francis turbine 49, 50, 51, 54, 56, 69, 108, 109, 113, 114, 121
  inlet 109
  system 50
Frictional power losses 201
Friction losses 44, 60, 79, 124, 137, 142, 144, 151, 152, 153
  estimation of 152, 153
  identical 79
Function 21, 24, 33, 86, 103, 139, 182, 183, 198
  hydraulic 21, 33, 86

G

Gear 32, 33, 74, 75, 76, 77, 78, 186, 187, 188
  driving 186, 188
  house 186
  identical intermeshing 186
  operating 32, 74, 75, 76, 77, 78
  regulating needle 33
Gear pumps 185, 186, 188, 202, 207
  external 186, 207
Gear wheels 186, 187
  meshing 186
Generation, peak power 160
Generator 29, 38, 114, 115
  efficiency 38
  electric 29
  poles 115
  rotor 114
Geometrical head 150
Geometric head 140
Gravity acceleration 140
Gross head 40

Guide 21, 32, 34, 50, 65, 74, 75, 88, 130
  blades 88, 130
  employed 74
  mechanisms 21, 50
  ribs 34
  sleeve 32
Guide vane(s) 50, 56, 63, 69, 74, 75, 76, 77, 83, 89, 90, 92, 96, 126
  angle 69, 89, 90, 92, 96
  canal 56
  width 63

H

Handwheele for spear adjustment 34
Head 115, 146, 168
  coefficient 146
  pressure 168
  ranging 115
Head loss 52, 53, 79, 80, 81, 98, 163, 178
  total 53
Height 51, 60, 63, 128, 140, 141, 169, 178
  atmospheric 178
  vertical 60
HEPP 33, 83, 84, 114, 115
  designing 84
  power house of 84, 114
  unit 114
High 21, 75, 126
  lift centrifugal pumps 126
  pressure oil 75
  water head applications 21
Hydraulic 4, 22, 88, 118, 123, 174, 193, 205
  brake 22
  energy 88
  head 118
  jump 4
  machinery 174
  oil 193
  pump efficiency 205
  turbines and hydroelectric power plants 123
  turbines classification 118
Hydraulic efficiency 42, 53, 65, 69, 91, 92, 95, 137
  theoretical 65
Hydroelectric 123
  power plants 123
  power stations 123
Hydropower 50, 121, 161
  global 121
  station 50

I

Identical 79, 81
  flow velocities dividing 79
  friction head losses 81
Impeller(s) 124, 126, 127, 128, 129, 130, 131, 132, 133, 134, 136, 138, 144, 148, 163, 169, 172, 185
  axis 134
  blades 144
  chemical cast steel 128
  construction 128Impeller element 185
  closed 128
  movable 131
  opened 128
  outlet 131
  passages 124
  single 126
  steed 128
Impeller diameter 140, 152, 153, 158
  nominal 158
Impulse turbine 11, 21, 33, 37, 86, 107
Inlet 65, 165
  angles 65
  blade angle 165
Installation 24, 50, 79, 129, 139, 142, 143
  conditions 129
  horizontal 24
  medium head 50
  vertical 24
Internal 34, 189
  gear pump 189
  turbine 34
Isothermal process 183

J

Jet 8, 16, 23, 33, 34, 35
circular 16
components 23
deflector 34
falling on moving curved plate 8
forces 35
free water 33
single 23

K

Kaplan runner design 64
Kaplan turbine 49, 50, 51, 55, 57, 64, 99, 100, 101, 107, 108, 114
runner 55
Kinetic energy 49, 57, 86, 124, 130

L

Law of Parallelogram of Force 8
Leakage 53, 136, 137, 138, 165, 201
internal 136
losses 165
Leather packing 34
Liquid 107, 124, 125, 126, 128, 129, 130, 131, 134, 136, 139, 140, 141, 144, 154, 155, 178, 186
column 178
efficiency 130
flow 131, 178
inters 129
outward 125
passing 186
path 131
pumped 141
velocity 130, 134, 140
viscous 128
Losses 40, 50, 53, 54, 56, 57, 59, 86, 88, 97, 107, 130, 132, 134, 136, 137, 139, 162, 178, 204
circulation 162
delivery pipes head 178
frictional power 204
glands 137
hydraulic 88, 136
lowest energy 56
mechanical 136, 137
pipeline 40
volumetric 54

M

Machines 1, 4, 103, 124, 125, 160, 161
hydraulic 1, 4, 103
Manometric 132, 134, 135, 137, 147, 163, 166, 173
efficiency 134, 137, 163, 166
head 134, 135, 147, 173
total 132
Mass of fluid acceleration 12
Mathematical 21, 24
calculations 21
manipulations 24
Mechanical efficiency 54, 137, 166, 201, 204
Meshing gears 187
Momentum 1, 2, 3, 4, 52, 124, 125, 136
change of 1, 4, 52, 124, 136
exchanges 2, 3
linear 1
Momentum equation 1, 52
linear 1
steady flow 52
Motion 1, 8, 11, 13, 14, 18, 19, 125, 176, 186
contrarotation 186
relative 1, 11, 13
Motor 160, 204, 205, 206, 208
hydraulic 204, 208
generator 160
input power 205
output power 206

N

Net 21, 124, 139, 140, 142, 143, 144
output power 21
positive suction head (NPSH) 124, 139, 140, 142, 143, 144
Nozzle 4, 6, 7, 21, 22, 23, 24, 29, 32, 33, 34, 38, 44, 46, 107
attached 24
small 22
straight flow 32
Nucleation growth 107

O

Open impeller pump 128
Outlet 2, 3, 8, 11, 13, 14, 18, 26, 27, 28, 30, 44, 47, 52, 56, 64, 90, 134, 166, 169, 172, 178
  angle 30, 169
  delivery 178
  flanges 134
  power 47
  shaft 44
  triangle 18
  velocities 11, 52

P

Pelton turbine 21, 22, 23, 25, 26, 29, 32, 33, 34, 35, 36, 41, 43, 46, 47, 106, 107
  components 36
  contraction 22
  single jet 29
  vertical 35
Penstock(s) 33, 50, 79, 80, 81, 82
  single 79, 81
  section 50
Piezometric head charge 118
Piston 32, 34, 75, 78, 174, 175, 176, 177, 178, 182, 195, 196, 197, 198, 203
  pump 177, 203
  servomotor 32, 34, 75
  stroke 195, 196, 198
  velocity 178
Power 1, 18, 19, 22, 43, 45, 46, 47, 52, 54, 61, 68, 83, 87, 100, 103, 104, 106, 107, 114, 115, 121, 125, 126, 136, 137, 146, 151, 152, 153, 154, 169, 188, 200, 201, 205, 207, 208
  absorbed 126
  brake 54, 200, 201
  consumption 151, 152, 153
  developed 38
  electric 154
  house 83, 87, 100, 114
  hydraulic 43, 45, 52, 125, 126, 207
  input 200, 205, 208
  internal 146
  mechanical 188
  output 136
  supplied 22
  theoretical 68
Power output 99, 207
  mechanical 207
Power plant 49, 83
  electric 49
Pressure 49, 86, 107, 108, 124, 126, 139, 140, 141, 183, 184, 185, 187, 197, 198, 200, 204, 205
  absolute 108, 139, 140, 141
  barometric 108
  initial 183, 184
  operating 200
  static 107, 187
  vaporization 139
Prototype hydraulic machine 103
Pump(s) 124, 130, 135, 147, 160, 162, 185, 186, 189, 200, 201, 204, 206, 208
  body 186, 189
  casing 130
  cavity flow 201
  flow rate 208
  head 162
  hydraulic 204
  impeller 1
  input power 206
  installation 135
  output 200
  power 126
  prototype 147
  rotary 185, 186
  turbines 124, 160
Pumped storage hydroelectric plants 160
Pumping 144, 150, 151, 205
  process 144
  systems 144, 150, 151, 205

R

Reaction force vector 3

Reaction turbine(s) 49, 50, 51, 52, 53, 54, 55, 57, 63, 65, 67, 71, 83, 84, 86, 95
axial flow 49
inward flow 49
water 84
Reciprocating pumps 124, 174, 175, 176, 177, 180
single acting 180
classification reciprocating 174
Rotary 185, 193
positive displacement pumps 185
vane pumps 193

S

Semi-open 128
impeller pump 128
type impeller 128
Servommeter piston 78
Servomotors 32, 75, 76, 77
application 75
cylindrical 77
hydraulic 32
Single penstock 80, 81
arrangement 80
installation 80, 81
Single stage centrifugal pump 127
Stroke 178, 179, 180, 203
Suction 108, 129, 131, 134, 135, 139, 140, 141, 142, 163, 164, 176 177, 178, 179, 182, 183, 193
head 139
inlet 131
pipe 129, 131, 135, 142, 163, 176, 182
pressure 108, 134
pump's 139, 140, 141
section 139, 140
stroke 179
system 139
tank 141
Suction lift 124, 142, 143, 163
and delivery pipes 163
negative 124, 142
positive 124, 143
Sugar molasses 128
System 1, 75, 79, 124, 139, 140, 183, 184, 204, 205, 208
hydraulic 124
regulating 75
suction piping 140

T

Teeth 187, 190, 202
adjacent 202
meshing 202
Temperature 107, 108, 128, 139, 141, 143, 144
function of 108, 139
hot water 128
lowering fluid 144
pumped 141
Theoretical pump torque 200
Theory 22, 52, 131, 175
of centrifugal pumps 131
of pelton turbine 22
of reaction turbine 52
of reciprocating pumps 175
Thoma's cavitation factor 112
Three-Screw Pump 191
Torque efficiency 204
Total hydraulic loss in turbine 53
Turbine 10, 21, 24, 32, 37, 38, 49, 50, 52, 75, 78, 84, 103, 104, 108, 109, 110, 112, 114, 115, 116, 117, 118, 119, 121, 124, 160, 161, 162
and pump blades 10
blade axial 115
cavitation in 103, 109
free jet 21
full-scale 110
high-head 78
high-power 75
high-power vertical 84
hydraulic 118, 119, 124, 160
mixed-flow 115
radial-axial 115
radial-axial flow 116
scale 110
speed 161
systems 50
Turbine runner 1, 11, 63, 66, 78, 86, 90, 98, 114
low-head 78
reaction 63, 90

V

Vane pump structure 193
Vaporization 144

Vapor pressure 59, 108, 139, 144
of liquid 139
Vapor tension 139, 141
Velocity 1, 3, 4, 6, 8, 9, 11, 14, 16, 18, 19, 23, 24, 29, 37, 39, 44, 49, 56, 57, 63, 64, 65, 67, 71, 87, 104, 131, 132, 135, 163, 169, 172, 178
absolute 8, 49, 56, 131, 132
and direction of water 169, 172
angular 11, 56, 57
bucket 29
coefficient 23, 29, 37, 44
constant 67
inlet outlet triangles 87
peripheral 9, 39, 71, 163
tangential 24, 63, 104
Vibration 107, 192
mechanical 107

W

Water 6, 8, 9, 13, 14, 19, 21, 32, 33, 42, 44, 46, 49, 50, 51, 60, 61, 67, 75, 84, 86, 87, 112, 113, 126, 128, 140, 141, 144, 160, 166, 168, 172, 175
absolute velocity of 8, 9, 166, 168
cold 141
dirty 175
hot 128
hydroelectric station 113
ordinary 128
power 42
pressure 75
pumping 160
quantity of 6, 21, 126
sewage 128
splashing 33
supply conduit 32
surface 144
turbine 87
Wheel 11, 24, 26, 37, 38, 39, 40, 42, 44, 46, 47, 64, 65, 67, 68, 71
diameters 40, 44
velocity 24

www.ingramcontent.com/pod-product-compliance
Lightning Source LLC
LaVergne TN
LVHW070117110826
845147LV00002B/140

* 9 7 8 9 8 1 1 4 9 4 1 1 6 *